Robert Young

An Essay on the Powers and Mechanism of Nature

Intended, by a deeper analysis of physical principles, to extend, improve, and more firmly establish, the grand superstructure of the Newtonian system

Robert Young

An Essay on the Powers and Mechanism of Nature
Intended, by a deeper analysis of physical principles, to extend, improve, and more firmly establish, the grand superstructure of the Newtonian system

ISBN/EAN: 9783337222291

Printed in Europe, USA, Canada, Australia, Japan

Cover: Foto ©berggeist007 / pixelio.de

More available books at **www.hansebooks.com**

AN
ESSAY
ON THE

POWERS AND MECHANISM

OF

NATURE;

INTENDED,

By a Deeper Analyfis of PHYSICAL PRINCIPLES,

To extend, improve, and more firmly eftablifh,

The GRAND SUPERSTRUCTURE of the

NEWTONIAN SYSTEM.

By *ROBERT YOUNG.*

LONDON:

PRINTED FOR THE AUTHOR,

By FRY and COUCHMAN, near Upper-Moorfields;

And fold by T. BECKET, Pall-Mall; J. JOHNSON, St. Paul's-Church-Yard; and J. MURRAY, Fleet-Street.

M DCC LXXXVIII.

PREFACE.

BY the Newtonian syftem, I mean the syftem of the world discovered by Pythagoras, revived by Copernicus, and, by the aid of a sublime and exquisite geometry, brought to its present state of perfection by Sir Isaac Newton. In this acceptation of the term, the following Essay is intended to support, extend, and improve the Newtonian syftem. But to do this, by no means implies an implicit attachment to every doctrine adopted in that syftem. On the contrary, improvements imply imperfection, and imperfection error. The physical parts of the Newtonian syftem are allowed, by its most learned and judicious friends, to want those traces of the great master that distinguish the mathematical principles.

Sir Isaac himself appeared sensible of their imperfection, and seemed to leave them to the correction of others, when he denominated his work mathematical principles, thereby leaving out the physical, as forming, in his opinion, a less perfect part of his system, on which he did not choose to commit his fame.

In reforming these, or attempting to do it, I think, far from opposing Sir Isaac, I am an humble coadjutor in his labours, by investigating, anew, those difficult points, in respect to which he candidly acknowledged his defects, and called upon the new exertions of his readers to supply them. Nor is it, truly, to honour the memory of this great man, with a superstitious veneration to receive, as sacred, all his opinions, and thus hand down his errors, uncorrected, to posterity, to the injury of his reputation when a greater lapse of time shall place mankind in a condition to discriminate more justly of his labours. Those, rather, may be called his

friends,

friends, who, by expunging what is erroneous in his works, leave that which remains more perfect; and by supplying the defects of his syftem, tranfmit it more complete to future times, a more glorious monument of his fame.

The following Effay is intended to improve the phyfical principles of philofophy, in order to extend our knowledge of nature. If I have fucceeded in thefe views it will be no objection to my attempt that I have deviated from principles already eftablifhed, fince it is obvious that every advance in knowledge implies the difcovery and renunciation of fome error. He, who will adhere immutably to tenets once adopted, can never increafe his light, nor improve his underftanding.

Nor will any, who think juftly, be offended, either that their own opinions are controverted, or that too little regard is paid to the authorities they have been accuftomed cuſtomed to reſpect. They will know, that while our knowledge remains imperfect, our opinions muſt be mingled with errors, and that thoſe errors muſt have the ſanction of the learned, who dictate opinions. They will pay the tribute due to the talents of every man, without yielding a ſlaviſh obedience to any. They will regard the inward conviction of the mind, as an authority ſuperior to the concurring ſentiments of the age; nor will they ſurrender up their own underſtanding to that of another, whatever be his talents or his name.

It is the more neceſſary to inſiſt on the liberty of reaſon, in matters of ſcience, when we find that liberty in danger from the zeal with which ſome endeavour to defend the opinions they have adopted; and when language and ſentiments like theſe are diſſeminated in philoſophical works;

" Sir Iſaac Newton is the perſon to whom
" we owe theſe obligations, and who is
" hence-

"henceforth to be confidered *as our only
"fure guide and inftructor;*" and again,
"Newton has difcovered the chaos, and
"feparated the light from the darknefs.
"His inimitable work, the mathematical
"principles of natural philofophy, con-
"tains *the true aftronomical faith: and
"thofe who reject its doctrines, are the
"worft of heretics*; as they fhut their
"eyes againft the cleareft of all light,
"demonftration[a]."

Surely, when implicit faith in any human authority is thus required of us, it is time we fhould difpute, if not becaufe we doubt, to affert, at leaft, the freedom of the mind, and preferve the right of doubting.

More congenial with the fpirit of fcience are fentiments, which, as delivered by a refpectable advocate for the principles of Sir Ifaac Newton, I avail my-

[a] Bonnycaftle's Introduction to Aftronomy.

felf

felf of, as a complete apology for my prefent undertaking, if any apology were wanting;

Mr. Maclaurin fays, " We often obtain
" this advantage from difputes concerning
" the elementary principles of any fcience,
" that they are the more carefully inquired
" into, and when found juft, are illuftrated,
" and *the better underftood for having been*
" *difputed*[b]." The fame gentleman, fpeaking of certain tenets, held by thofe who oppofed the Newtonian fyftem, fays, " If
" thofe tenets be true, they will be con-
" firmed by our inquiries; and if they
" be falfe, furely, it is better they fhould
" be detected[c]."

And is not the fame to be faid of all principles and doctrines? If they be true, they will be confirmed by our inquiries; and if falfe, furely, it is better their fallacy fhould be detected.

[b] Maclaurin's Account of Sir ISAAC NEWTON's Difcoveries, page 151.
[c] Page 5.

It

[ix]

It is acknowledged by all, that much obfcurity and difficulty hath ever attended the fubject of the following difquifitions. It is by freedom of inveftigation alone, that we can hope to make advances in the inquiry. The voyager, who confines himfelf to a long accuftomed tract, can have no profpect of making difcoveries. I hope others will follow my example, and that this important fubject will be brought forward into a full difcuffion. And if I fhall be found to have erred in my attempt, I fhall yet be confeffed to have rendered fome fervice to fcience, if what I have done fhould be the means of inciting others to more fuccefsful endeavours.

As, in practice, nothing is perfect, and few things wholly without merit; fo, in theories, perhaps none are without error, nor any devoid of truth. The difference between opinions, feems to lay chiefly in the different proportions of truth and error which they contain. If this be true, every advance in principles is only fubftituting

a   a lefs

a lefs imperfect theory for one more fo, and the laft, ever leaves fomething for futurity to correct.

Some readers may be defirous to find a name for the following work, by obferving to what fyftem, ancient or modern, it moft adheres, or is moft refembled: for myfelf, I can affure them I did not feek for it, nor find it, in books, but in the reflections of my own mind alone.

Many will object to the method of reafoning herein followed, that philofophy has only to collect general laws from phenomena, and then apply thofe laws to the explanation of other phenomena; and has no concern with metaphyfics. To this I anfwer, that the prefent is a phyfical, and not a metaphyfical work, if metaphyfics *are* to be loaded with an unmerited opprobium; that the object of my inquiry is not facts but powers, not effects but caufes; that the received theory doth not, agreeably to its profeffions, confine itfelf

to

to the affertion of phenomena, but affumes, as caufes, phyfical principles, which are not matters of fact, nor laws collected by induction, but which are to be examined by phyfical reafoning alone. Thefe principles, thus affumed in the received theory I have not adopted, and therefore it became neceffary to fubftitute others in their place, and to this end I have employed the following pages. I have not attempted it by means of experiments, becaufe they can difcover effects only, and not caufes, which I fought; nor by geometry, becaufe it is concerned only with relations of quantity, and cannot lead to the knowledge of being or of power.

By phyfical reafoning alone, I have found a phyfical principle, adequate to the purpofes of explaining phenomena. A fubftance actually exifting, poffeffed of active powers, the bafis of matter itfelf, and the agent in all effects. This active fubftance appears to have been the defideratum in all ages of philofophy. It removes

removes the obscurity which attends the consideration of matter as an original and inactive substance, whose essence, solidity, is confessed to be incomprehensible; it removes the difficulty which has ever attended the question of the origin of motion, by showing motion to be the original form of being; and thus reflects a light upon the very foundations of science.

If, in this pursuit, I have been obliged to throw down some barriers of ancient opinions, I hope the success will repay the sacrifice. That matter was an inactive and impenetrable essence, was an error of early date. Led away by this, Sir Isaac Newton was obliged to frame such principles as were agreeable thereto, and the second error was a consequence of the first. He ascribed to matter a quality, till his time unthought of, an inherent power of persevering in its proper state of rest, or uniform rectilineal motion. A principle, the importance, novelty, and singularity of which, entitled it to a more cautious

scrutiny

scrutiny than it seems to have met with. It *begged the question*, long disputed, by what cause motion was continued, averring, but without proof, that it was by a force innate in bodies. This principle being only intended to explain the perseverance in one state, required another to account for changes of the state of bodies from motion to rest, or from rest to motion.

A name was given to this, *impressed force*, but no explanation given of its nature, seat, or origin. The objections to these principles are many and unanswerable. They are wholly destitute of proof; they are far from being evident in themselves, otherwise, how were they so long unknown. They present no clear idea to the mind, and want that precision so requisite in first principles. The *vis inertiæ* has a contrariety of nature that perplexes the understanding. It has no decided character, is neither motive nor quiescent; but alternately and indifferently both; by being

being too much, it becomes nothing; by uniting two contraries it deſtroys both. The *vis impreſſa* hath no clear idea annexed to it, neither as a ſubſtance nor a quality; it hath no intelligible mode of operating; it acts, and ceaſes to act, in a manner equally inconceivable.

Wanting ſimplicity, clearneſs, and conſiſtency, theſe principles afford very little ſatisfaction to the mind; wanting proof or evidence, they have no other claim to our acceptance, than the authority of their author, which hath too long, already, preſerved their credit with the world. So long as we conceived matter to be impenetrable, and endowed with a power of perſevering at reſt or in motion, it appeared to me, that no true or ſatisfactory principle could be diſcovered for the explanation of effects. It was neceſſary to get rid of theſe before others could be ſubſtituted in their room.

For this reaſon I publiſhed, ſome time ago, an examination of the phyſical principles

ciples of the received fyftem[d], wherein I showed their error, independent of any view to fubftituting another theory in their ftead. Their error was evident from the contradictions they appeared to involve; they were shown to be infufficient to explain phenomena; and it is, then, of little confequence to allege their defect of proof. It was impoffible, in this matter, to pay regard to the prevailing prejudices of the day, without facrificing the more important interefts of truth. By showing the infufficiency of received principles, I intended to open the way to new difquifitions upon the fubject; and fuch an one I here fubmit to the learned.

Had I delayed the prefent work fome time longer, I might have corrected many of its faults, and confulted my own reputation more; but other avocations required me to difmifs the prefent: and I hoped this great advantage would arife from its early

[d] Entitled, An Examination of the Third and Fourth Definitions of Sir ISAAC NEWTON's Principia, and of his Three Laws of Motion.

publication,

publication, that others would be the sooner induced to profecute fo important and extenfive an inquiry, in the fame method. I have chiefly confined myfelf, in this volume, to the inveftigation of general principles, and hope, in a future work, to profecute further, fome particular applications to phenomena.

On a fubject fo difficult, treated in a method in which I had no guide nor affiftance from others, I hope much indulgence will be granted me, and I am confcious that I ftand much in need of indulgence. I requeft that the reader will diftinguifh between the defects of the author's abilities and the faults of his principles. Many truths he may find not explained in the beft poffible manner, nor fupported by the moft cogent proofs. In fupplying fuch defects, and correcting what errors I may have fallen into, there will be fufficient exercife for the talents, and the good nature of thofe who choofe to undertake thefe tafks.

DEFINITIONS.

DEFINITIONS.

ACTION. THE production, or prevention of any change.

ACTIVITY. Disposition to motion.

AGENT. A being which acts.

ATTRIBUTE. That which exists in another and not of itself.

CAUSE. That which produceth an effect.

EFFECT. That which is produced by another.

MATTER. A being whose parts resist penetration or separation.

MOTION. Change of place.

POWER. Ability to act.

RESISTENCE. Counter action.

CONTENTS.

PART I.

Analysis of Matter and Motion.

INTRODUCTION................Page 1

CHAP. I. *Analysis of Matter in general*......p. 8

CHAP. II. *Analysis of Atoms, or Primary Parts,* p. 23

CHAP. III. *Error of the Doctrine of Impenetrability,* p. 31

CHAP. IV. *Analysis of Motion*............p. 45

CHAP. V. *Action is an Attribute of an Active Substance*...............................p. 57

CHAP. VI. *On the Existence of an external, unintelligent Agent, with a Review of several Opinions,*

Opinions, particularly that of BERKLEY, on the Subject ....p. 63

CHAP. VII. The ACTIVE SUBSTANCE an immaterial Existence, not Mind, but intermediate, and related to Matter and Mind ....p. 84

PART II.

Of Action, and the Manner in which the ACTIVE SUBSTANCE *produces Matter and Motion, investigated by Inferences from Effects to Causes.*

INTRODUCTION..............Page 89

CHAP. I. *On the Rules of Reasoning from Effects to their Cause*............................ p. 91

CHAP. II. *In what Manner the* ACTIVE SUBSTANCE *is active, and how it acts*........p. 104

CHAP. III. *On the different Origins and Circumstances of Activity in Bodies, apparent from Observation* ....p. 108

CHAP. IV.

CHAP. IV. *Of Activity in the General, considered as appertaining to Bodies* .... p. 113

CHAP. V. *Of Activity in Bodies, as distinguished into Motion and the two Forms of Impulse, Pressure and Percussion* .... p. 122

CHAP. VI. *Of the Production of Matter from its immaterial Elements* .... p. 147

CHAP. VII. *Of the Constructive Analogy between the Primary Corpuscles, the Planets with their Systems, the Solar System, and the Fixed Stars,* p. 193

PART III.

A further Investigation of the Nature and Laws of the ACTIVE *or* ELEMENTARY SUBSTANCE.

CHAP. I. *Origin of the Physical Power, exemplified in Elastic Bodies, Inflammable Bodies, Chemical Operations, Living Processes, and Gravity* .... Page 215

CHAP. II.

CONTENTS.

CHAP. II. *Origin of Motion* .... p. 222

CHAP. III. *Origin of the Orbicular Motions, which the* ACTIVE *or* ELEMENTARY SUBSTANCE *assumes, as the constructive Principle of Nature; and the Laws of the Elementary Substance,* p. 233

CHAP. IV. *The Mechanism of Cohesion, and the connecting Corpuscles* .... p. 247.

CHAP. V. *Of the Orbicular Motion of the* ACTIVE SUBSTANCE *in Mechanical Impulse* ...... p. 257

CHAP. VI. *Absurdity implied in the Question concerning the Origin of Motion* .... p. 263

CHAP. VII. *Law of Union, in what Respects agreeing with the Attraction of the Newtonian Philosophy, and wherein distinguished from it,* p. 268

PART

PART IV.

Some Abridgement of the foregoing Ideas; their Agreement with Facts: Conclusion.

CHAP. I. *Constitution of the Universe*, Page 273

CHAP. II. *Agreement of Phenomena with the Theory*..............................p. 283

CHAP. III. *Agreement of Phenomena, with the Orbicular or Revolving Motion, of the* ACTIVE SUBSTANCE, *which is the Constructive Principle throughout Nature*................. p. 308

CHAP. IV. *Select Propositions, abridged from the foregoing Pages, and separated from their Proofs,* p. 322

CONCLUSION......................... p. 331

ERRATA.

Page 46, *line* 12, *for* in, *read* is.
—— 58, — 10, *for* action, *read* to act.
—— 150, — 7, *for* and, *read* or.
—— 291, — 17, *for* action, *read* re-action.
—— 306, — 4, *for* double, *read* quadruple.

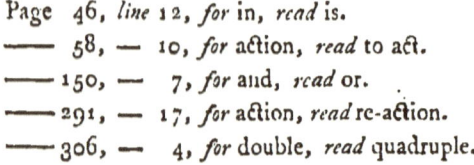

I beg leave, in this place, to correct an error in my former work, *An Examination of the Third and Fourth Definitions of Sir* ISAAC NEWTON's *Principia, &c.* where, speaking of a horse drawing a stone, I say, " The stone doth not re-act, because " it doth not act; it resists, but resistence is not action." Page 31.

It is usual with authors to ascribe the difficulty of moving bodies to a resistence of the bodies; and I incautiously, from habit, fell into this error, although I had, before, corrected it in myself. My mistake lay, first, in admitting that bodies, at rest, independent of friction or media, *resist*; and I was then led to the inconsistency of denying resistence to be action. But the difficulty experienced in moving a body, doth not proceed from a resistence of the body: a body is not moved without an adequate cause; and the difficulty lies in imparting that cause to the inactive matter, not in any resistence it opposes to being moved.

AN

ESSAY, &c.

PART I.

Analysis of Matter and Motion.

INTRODUCTION.

THE first part of the business of Philosophy is to seek for principles, the next, is to apply them to the explanation of natural appearances. Principles are distinguishable into two kinds, EXPERIMENTAL and RATIONAL; The experimental, are general facts, which, being found uniformly to obtain, so far as observation has gone, are confided in, as constant, and referred to as DATA, by which to explain other less general facts which they involve. Rational principles are conceptions of the understanding, whose evidence rests not alone on experiment, but on intuitive perceptions.

[2]

B   To obtain the former fort of principles, we muft make obfervations of facts; to arrive at the latter, we muft exercife our intellectual faculties.

C   Powers and primary caufes are objects of the underftanding alone. Some have erroneoufly conceived, that ultimate caufes lay concealed in the minute parts of matter, and efcaped us only by their diminifhed magnitude, and imagined that finer organs are all that is wanting to their difcovery. But this is not the cafe: ultimate principles cannot be fubjected to the examination of our fenfes, becaufe every perception of fenfe is an effect of thofe principles, and prefuppofes them. Thefe, therefore, are not to be fought by experiment, but inveftigated by reafon.

D   Among rational principles, or truths intuitively perceived, we are not to choofe at random or by guefs, fome one for the bafis of a philofophical fyftem. For out of a multitude of truths, equally certain, how fhall we afcertain thofe which will ferve to explain the greater number of phenomena?

E   Des Cartes appears to have erred in the method he took to eftablifh his rational principles. He
fought,

fought, among the number of truths which arofe to his mind, one which fhould be primitive, and involve all the reft. But in this he undertook a tafk far too great, at leaft for the then or the prefent ftate of fcience. He firft perfuaded himfelf that there was but one primitive truth; he then concluded that truth could be no other than the certainty he had that he thought: but he was unqueftionably miftaken. We are as certain of our paffions, and of our fenfations, as of our thoughts; and from each and every confcioufnefs we know our own exiftence, which he propofed as deducible from thought alone.

His firft principle, " I THINK," although no one will conteft its truth, had no particular relation to the phenomena of nature, more than any other fimple truth, I feel, I fee, I tafte. It was a truth far too general, and too remote from material phenomena, ever to lead him to a rational explanation of them. Accordingly, we find no clofe and connected chain of reafoning follows it, but mere affertion or hypothefis.

All our knowledge, of every kind, originates in our external fenfations: for it is this fource which furnifhes us with the occafions of our internal feelings or paffions; and both of thefe,

external

external impreffions, and internal emotions, afford us ideas, about which our thoughts and underftanding are employed.

H It is, therefore, from this fource, external fenfations, that we are ultimately to derive, alike our rational and our experimental principles.

I The material world, and the fenfible appearances around us, are what we feek to explain. Where then fhould we look for firft principles to guide us, but in thofe things we propofe to develop. To begin in our own thoughts, in order to difcover the caufe of an external appearance, feems as little reafonable, as if we fhould confult the cuftoms and opinions of the Chinefe to account for a tranfaction in modern Europe.

K We have, therefore, confidered the world as it is manifefted to us by our various organs, and impreffes us with the firft rudiments of knowledge, as being the only fource from whence we are to endeavour to obtain the firft principles of philofophy, both rational and experimental. And firft we propofe to inveftigate the rational principles.

The

[5]

The principles of every thing are fought by an analyfis, which, in the compound forms, difcovers the fimple and component parts. Chemiftry analyzes compound bodies to arrive at their principles. Natural Philofophy, in like manner, analyzes the compound appearances of motion, and refolves them into fimple: thus Newton's laws of motion have been formed. We muft, if we would arrive at true rational principles, feek them in the fame manner, by a refolution of the compound in which thofe principles are contained.

The material world, or fenfible appearances, contain, we faid, the principles of all knowledge. It is thefe, therefore, in which we are by various methods of analyfis, to feek the different forts of principles.

The Chemift, by a feparation of diffimilar parts, gets nearer to the primitive forms of bodies. The natural philofopher, purfuing the fame means with motion, approaches nearer to its original laws: thefe both proceed by experiment. We, as we feek not either motion or matter, but in matter and motion to find an idea yet more fimple, and which fhall be prior to them and involve them, muft quit the method of experiment, and commit ourfelves to our reafon.

But

o But let none imagine we shall therefore proceed with the less certainty, or that we must forsake all confidence in ourselves, as soon as we leave the limits of sense: neither, in fact, shall we have occasion to go out of the limits of sense; for although we are, in this inquiry, to employ our reason without the aid of experiment, yet we shall employ it only about facts and sensible things; and whatever we shall perceive by a rational analysis, that sensible impressions contain, as their ultimate principles, will be no less certain than those things which experiment teaches us; since both alike, our thoughts, and our sensations, afford us certainty, and serve as the basis of reason and knowledge.

p Facts, as perceived by the unlearned, are a kind of middle point, from which two classes of philosophers proceed in contrary directions. The experimentalist, first analyzing these compound appearances, and discovering the principles or laws of motion, then proceeds to combine and to explain the same appearances, by his method of synthesis. But he, as his causes or first principles, go no farther back than simple facts, is more properly to be called the natural historian of motion.

q The other class of philosophers, not merely
seeking

seeking amongst sensible appearances, what are the most simple, but, further, seeking in those simple sensible appearances, for an idea yet more simple, which, not being discoverable by sense, is yet evidently the origin and source of sensation; he is truly investigating causes, and this is more strictly to be termed philosophy. The higher principles, found by this analysis, will serve to involve and explain those facts, which, to the experimentalist, are simple and primary. The first experimental principles, are no longer the first principles of philosophy, but are deducible from prior ones, which this more sublime analysis has led us to discover. This continues the chain of human reason farther back into the recesses of nature, and brings us nearer to the first cause.

Both of these paths of inquiry proceeding from one middle point, sensation, and meeting there, form a connected whole, whose parts mutually conspire to its perfection, and to each others support.

CHAP. I.

Analysis of Matter in general.

A   WE are to seek to arrive at first principles, which both truly exist, and are truly the origin of those sensible facts philosophy endeavours to explain.

B   These principles we shall find, by resolving the complex ideas of bodies and motions into their constituent and primitive essences.

C   The nature of bodies, signifies the aggregate of all those ideas with which they furnish us, and by which they are made known to us.

D   Bodies have various relations to us, each of which is the nature of the body, as relative to our nature, and the sum of these relations is the whole relative nature of the body. The absolute nature of body is not to be sought among its sensible qualities, since all of these are its relations to us; and in these relations, it would be in vain to expect to find that which is not relative, but absolute.

We

We become acquainted with bodies through all the different mediums of our senses; our sight, our touch, our hearing, our taste, and our smell. Each of these mediums affords us, with its own peculiarities, a great variety of sensations: our sight, figures and colours; our touch, hardness, softness, roughness, smoothness, heat, cold, moisture, dryness; our hearing, all the variety of sounds; our taste, of flavours; and our smell, of odours.

All the ideas with which any one body furnishes us, make up the whole nature of that body; and all the ideas which we get from all the varieties of bodies, make up the sum of our knowledge of the nature of all bodies.

Every body furnishes us with a number of ideas: as, for example, figure, size, weight, hardness, or softness, colour, taste, motion, or rest, and others, may all present themselves in one body.

Although these are linked together in the object, they are distinct in the mind, and can each be considered apart from any other.

The whole compound, in which these ideas are united, we call the body; each of these ideas,

ideas, confidered apart from the reft, we call a quality of the body.

K   Among thofe qualities fome are particular, fome more general, fome univerfal. To attract iron is the quality only of the loadftone or magnet; to ferve for refpiration only of air; to afford nutriment, of moft animal and many vegetable fubftances; to be extended, figured, folid, and moveable, of all bodies.

L   The qualities of bodies are diftinguifhed into two claffes: of one it is faid the ideas in our minds, which we call qualities, have their fimilar archetypes in the bodies, and thefe are called primary qualities. Extenfion, figure, folidity, inactivity, and mobility, are called primary qualities, being fuppofed to be truly in bodies, juft as they are reprefented in the mind.

M   The other fort of qualities are confidered as not being in the bodies, but as being only the fenfations produced in us by the action of the primary qualities on our fenfes; fuch are heat, cold, tafte, colour, hardnefs, found, and odour.

N   In analyzing body we may, therefore, pafs over thofe qualities which have no exiftence but in our minds; fince we cannot find in bodies

colour,

colour, nor sweetness, nor pain, nor sound, these will have no place in our analysis, not being present in the thing we analyze, nor making any part of it.

There is, however, in the bodies which furnish us with the ideas of the qualities called secondary, something whereby they are able to produce those ideas in us; for though the ideas are only within us, they do not arise in us of themselves, but have an origin in the bodies.

O

We do not know what it is in the bodies which is able to produce in us ideas unlike to itself; but we express the faculty of doing this by the term power, and say bodies have a power producing in us ideas of sweetness, yellowness, heat, &c.[a]

P

Hence in the body itself, we are, in the place of the names of all the secondary qualities, whose ideas are in our minds, to substitute one idea of power, and to conceive that the power to produce those changes in us is in the body.

Q

Body, then, may be said to contain, or consist of, all its primary qualities, extension, solidity,

R

---

[a] Mr. Locke refers secondary qualities to powers in the bodies.

C 2   figure,

figure, inactivity, and mobility, together with a power to produce certain effects on our bodies; to which we may add a power to produce, also, certain effects on other bodies, as the power of aqua regia to diffolve gold, and fire to melt wax.

s   This divifion of bodies into feveral component parts of one united whole, is itfelf an analyfis already made; and the enumeration of the primary qualities of body, together with the power to produce the fecondary ones, may be called the fimple parts, which arife from the mental decompofition of the compound idea of body.

t   But it may be inquired, if this analyfis which others have made, be both juft and perfect; or if, being imperfect, it may be improved and carried to a greater extent.

u   Let us confider firft the primary qualities of bodies, extenfion, figure, folidity, inactivity, and mobility, and fee if we can reduce them to any more fimple parts.

x   EXTENSION feems to be a fimple idea formed only of parts like itfelf; every extended being or fpace is compofed of lefs extenfions, ad infinitum, fince no bounds can be fet to divifibility.

FIGURE

FIGURE is merely the limit of a finite extenfion. Extenfion may be carried on all fides equally diftant from a centre, making a fphere; or fome parts may be more extended than others, making any angular figure: figure is nothing in the nature of bodies, but a mere confequence of extenfion: thefe two, therefore, extenfion and figure, afford little matter for analyfis.

SOLIDITY is the quality of body which principally requires our notice; it is that which fills extenfion, and which refifts other folids occupying the place it occupies, thus making extenfion and figure fomething real, and different from mere fpace or vacuity. If the fecondary qualities of bodies, or their powers varioufly to affect our fenfes depend on their primary qualities, it is chiefly on this of folidity; for without hardnefs and refiftence, fize and figure would do nothing towards making any impreffion, or producing any effect. Solidity is therefore the moft important of the primary qualities, and in it the effence of body is by fome conceived to confift.

This idea of folidity has been judged to be incapable of any analyfis, and not to admit of being carried beyond the intelligence which we receive of it from our fenfes. Notwithftanding standing this, it appears evident to me, that the idea of solidity may be resolved into another idea, which is that of power, and that solidity depends as much on power as do the secondary qualities.

B   We can only conceive of solidity as being a resistance of the parts of any body, to a power which endeavours to separate them, or to bring them nearer together. Now that which resists any power, and prevents its effect, is also a power: by resistence I mean here an active resistence, such as an animal can employ against an animal. If a horse pulls against a load he draws it along, but if he pulls against another horse he is put to a stand, and his endeavour is defeated. When any endeavour to change the situation of the parts of any solid is in like manner prevented from taking effect, and the parts retain their situation, the situation has plainly been preserved by an active resistence or power, equivalent to that which was fruitlessly exerted on them.

C   Since then the solidity of body does necessarily imply an active power of resisting within the extension of the body, it becomes unnecessary, and even inadmissible, to suppose that the solidity folidity in the body, is at all a pattern or archetype of our fenfation.

Philofophers do not allow colours, taftes, or founds, to be in the bodies, becaufe thefe ideas are plainly produced in us by the mediation of our organs. There appears to me an incongruity in fuppofing folidity is any more in bodies than colours and flavours are, fince it is equally with them a fenfation, and an idea to which the mediation of our organs is neceffary. Mr. Locke fays, " If any one afks me what this folidity is, I fend him to his fenfes to inform him." The folidity then is in our fenfes, not in the body. If any one afks what a blue colour, or a fhrill found is, he will alfo be fent to his fenfes for information; each is an idea of fenfe exifting only in it: and it appears to me as little juftifiable to fay folidity is in a body, as to fay heat is in the fire; and this latter Mr. Locke thinks as improper, as to fay, pain is in the knife that cuts a perfon.

If it is faid, it is not the fenfation of folidity which is in the bodies, but a real folidity, of which the fenfation is the pattern; I reply, if the folidity in the body be a perfect pattern of the fenfation, it is a fenfation; but this cannot be admitted: if it be any thing different from

the

the fenfation, what is that difference? This we cannot learn from our fenfes.

F   Solidity as it really is (or is faid to be) in bodies, is confeffed to be utterly incomprehenfible. The fame author fays, fpeaking of the parts of water cohering into ice, " He that could find the bonds that tie thefe heaps of loofe little bodies together fo firmly; he that could make known the cement that makes them ftick fo faft one to another, would difcover a great, and yet unknown fecret; and yet, after that was done, he would be far enough from making the extenfion of body which is the cohefion of its folid parts, intelligible, till he could fhew wherein confifted the union, or confolidation of the parts of thofe bonds, or of that cement, or of the leaft particle of matter that exifts; whereby it appears, that this primary and fuppofed obvious quality of body will be found, when examined, to be as incomprehenfible as any thing belonging to our minds, and a folid extended fubftance as hard to be conceived as a thinking immaterial one."

G   If folidity as a quality in bodies is incomprehenfible, it is to little purpofe to infift on its actual prefence; and if we can conceive of matter as well without this myfterious quality, it
cannot

cannot be desirable to retain it. As a sensation in ourselves, we can perfectly comprehend solidity; in body it is altogether unnecessary; for a power of active resistence only is required to the making impressions on our senses, and to all the qualities which body possesses.

Solidity, confessed to be itself incomprehensible, cannot serve to render body more intelligible; on the contrary, it must necessarily carry along with it its own obscurity, and perhaps, it will be found, that to it alone is owing all the difficulty men have experienced in developing the essence of matter.

Since then, solidity, as a sensation, can exist only in our feelings and ideas, and as a cause of the sensation existing in body can only be conceived to be an action, all our endeavours to form any other idea of it, being ineffectual, we ought to consider this quality as an effect only, and not to imagine that any resemblance of the sensation exists in bodies, nor that bodies are in any other sense solid, than as having a power to resist, from which power our sensations of solidity and hardness arises.

INACTIVITY is the next primary quality we are to examine. Our having referred solidity to an action in bodies, thereby leading to the conclusion that matter is effentially active, may feem incompatible with the idea of its being inert.

L   But the activity of matter whereby it refifts and gives us the idea of folidity, is to be confidered in a certain limited fenfe, and its inertnefs, of which we now fpeak, is to be regarded in another limited fenfe, and both of thefe are compatible within their refpective limits.

M   The activity of body may be confidered as belonging to the parts of a compound; its inertia, as the inertia of the unit formed of thofe parts: the actions of the parts are every where oppofed to each other, and equal; from hence refults the inactivity of the whole: this will be explained more at large in the fequel.

N   The inactivity of matter is a privation, and affords no fubject for analyfis.

O   We ufe the term inactivity in its fimple and vulgar fenfe, and not as it is ufed in the Newtonian philofophy; it is there confounded with action, and the diftinction between activity and inactivity is deftroyed, becaufe inactivity is made common and indifferent to motion and reft. Now if motion be derived from inactivity, what

left

teſt have we of action, or how can we define the term? It is ſuppoſed that the continuance of a body in an uniform rectilinear motion is no change, and ought, therefore, to be a conſequence of inactivity; but this ſuppoſition is erroneous, and the conclusion from it is repugnant to the nature of things (b).

MOBILITY is the laſt of the primary qualities commonly aſcribed to body; this is only a capacity of receiving motion, and no poſitive idea, and therefore affords no materials for analyſis.

Thus we have conſidered diſtinctly all the ideas which we derive from body, as they are commonly enumerated and claſſed; and in analyzing all the ideas we have of body we have analyzed body itſelf, ſo far as it is known to us.

Of its primary qualities, EXTENSION, is not peculiar to body, but belongs equally to the idea of ſpace: extenſion therefore, conſtitutes nothing of the nature of body; FIGURE is a conſequence of finite extenſion; INACTIVITY is only a privation of a motive tendency; MOBILITY is a capacity of receiving motion; and none of theſe can be rendered more clear or ſimple by analyſis.

ᵇ See Examination of Third and Fourth Definitions.

s   SOLIDITY, alone, of the primary qualities, is pofitive and peculiar to body, and this, on being analyzed, refolves itfelf into ACTION or POWER: all the fecondary qualities are generally admitted to be powers.

T   Thus all which is real, pofitive, and peculiar to body, are certain active powers, which varioufly affect our different organs, and produce in us all the ideas of fenfe.

U   The effence of body is then, by our analyfis, reduced to power: what can we yet further fay of the idea of power? If folidity is found not to be in bodies, but to be an idea produced in us by a power in bodies, is the power then really in the body, is it without us, and independent on our fenfations?

X   The idea of power is not acquired by our fenfes, and is not a fenfation as folidity is. We deny folidity to be in bodies, becaufe it is a fenfation; bodies have not the fenfation, and the term folidity, if not applied to the fenfation, has no meaning but what is more clearly expreffed by power; therefore we fubftitute power for it, as it is referred to actual exiftence in the bodies.

Y   Power is an idea of reflection, it is fuggefted

to

to us by thought; now if we know any thing of real exiftence in body, our fenfations, derived from body, not being that real exiftence, but effects thereof; it follows that our knowledge of the real exiftence in body muft be fuch as is fuggefted to us by our thoughts, exercifed about our fenfations.

Our own being furnifhes us with the original idea of power. We are capable of acting and producing changes in appearances, and this faculty which we experience to exift, we call power. We do not indeed know what is the effence or origin of the power, nor how it effects the change; our idea of it is therefore imperfect; it is that which can produce a change or can act. This is its characteriftic, by which it is diftinguifhed from all other exiftences which are incapable to act, or produce change.

We are confcious of the exertion of our own power; when, therefore, we fee ACTION or CHANGE happen without any exertion of power on our part, we refer this to other powers, without us, and necelfarily conclude the POWER to exift where the CHANGE begins, or the ACTION is exerted.

This power, then, referred to bodies, muft exift in them, or it can exift no where. It does
not

not exist in our minds; in our minds it is only an idea of a resemblance to the power we are conscious of in ourselves, associated with those external appearances where we experience changes and actions. The idea in our minds is, therefore, a fallacy if the thing does not exist without the mind. From hence we conclude that power truly exists in bodies.

CHAP.

CHAP. II.

Analysis of Atoms, or Primary Parts.

ALL bodies, of a senfible magnitude, are A suppofed to be porous, or to have spaces interfecting their internal continuity, either void, or filled with a lefs refifting fubftance.

Thefe pores, which form, within a mafs, a B number of diftinct furfaces, lead naturally to an ideal divifion of it, by an entire feparation of the parts, which the pores, in fome meafure, detach from each other.

It is eafy to conceive that the parts, which, by C their affemblage, and by touching only in fome points, form fenfible and porous maffes, may be themfelves porous, or made up, each, of fmaller parts which touch only in certain points, and which lead in like manner to a divifion of them; thefe pores will be pores of a fecond order, and the parts, parts of a fecond order: the fame idea may be continued through any feries of orders of pores and parts.

But

D   But fuch a feries muft ultimately terminate in parts which are not porous, otherwife the body would be all pore, which is impoffible.

E   These ultimate parts, void of pore, we muft confider as containing the effence of a body, becaufe all that part of the fenfible magnitude which confifts of interftices between the folid parts, is not body.

F   In any hard or refifting fenfible mafs, formed of an aggregate of thefe ultimate parts, any how put together, and arranged into any number of different orders of pores and porous parts, two things are to be confidered as neceffary to the hardnefs or refiftence of the fenfible mafs.

G   Firft. Hardnefs or refiftence in every ultimate part itfelf; for if each were not hard in itfelf, their affemblage could not conftitute a hard mafs.

H   Secondly. Where thefe hard parts touch, they muft not only touch, but prefs together, fo as to refift feparation; for without this there could be no union among them, nor would their affemblage form a whole or united body.

This

This preſſure together of the parts of bodies is called coheſion.

It is common to conceive of coheſion as a power which ſerves to cement together the ſolid parts of bodies, operating at the ſurfaces of the ſolid parts, and not to conſider it as belonging to each ſolid part ſeparately, exiſting within it, and cementing together its own parts.

But diviſibility does not terminate where poroſity ends; the ſolid parts, which are, in regard to the orders of pores, ultimate parts, are yet extended maſſes: they have magnitude, and thoſe magnitudes have their parts diviſible, ad infinitum.

If a ſolid part, void of pore, is an united whole, it is becauſe its two halves are attached together by a power which prevents their ſeparation, and becauſe each half has its halves, in like manner, cemented.

The attachment of any two parts of a ſingle corpuſcle is of the ſame nature, and can be conceived of no otherwiſe, than as the attachment between any two or more corpuſcles which form a ſenſible maſs.

o   For any two corpuscles, when cohering, form a single mass, of which each corpuscle may be conceived a half; the cohesion of these can only be conceived of as a power, holding them together by the touching surfaces.

p   Every corpuscle has its two halves, or may be conceived as divided into any two portions, whose surfaces coincide, in like manner as the coincident surfaces of two corpuscles which form a single mass; and these surfaces of the two portions of a single corpuscle, are held together by some power which resists their separation.

q   And the power in the case of the two parts of one corpuscle, or of two corpuscles, can be no ways different; power being always one simple idea, like to itself: two corpuscles form one mass, just as the two halves of a corpuscle form one corpuscle.

r   We may conceive the power to exist in different degrees. We may conceive two parts of the same corpuscle, held together more strongly than two corpuscles; and that any two corpuscles would separate with a less force than would be required to break either of them; but these are different degrees of the same thing.

We

We are not, then, to conceive of cohesion as s a power existing only without the primary corpuscles, and operating only at their surfaces to attach them to each other, but also as a power whereby the parts of each corpuscle are held together, and by which the corpuscle exists, a solid and extended mass.

Nor can these primary atoms consistently be T denied to have parts, or be divisible; although they may have no division by pores, nor are ever in the order of nature broken down into smaller parts, and are thus, in respect to the actual course of events, primary and indivisible atoms, yet every atom has magnitude, and consequently parts, which, though never separated from each other, exist distinct in the whole.

But as these parts are distinct, they are separable, and are held together by some bond of union or cohesive power. U

Every part of a primary corpuscle may be X considered as attached or cohering by its surface to a contiguous part of the same corpuscle.

These contiguous and cohering surfaces may Y be assumed every where within the corpuscle.

E 2   At

z   At every surface nothing is found but an attaching or cohering power; for the surfaces themselves are not parts of the corpuscle, but terminations of the distinct parts.

A   Therefore, on pursuing a division of the corpuscle indefinitely, power is the idea which continually occurs, and the resolution of the idea of the corpuscle continually into smaller parts, presents to us continually the powers which mutually attach those parts; nor can any other idea be found.

B   If it be said, that by thus dividing a corpuscle in idea, we get a continued division of *solid parts*, held together by *power*, and so preserve the idea of solid parts cohering, I ask, what is this idea of solid parts, which we have distinct from the powers by which they cohere; what is there in the solid mass, besides the power which makes it hard or solid?

C   This we cannot learn; we have no idea of hardness or solidity but by our senses, and these sensations are not in the body, but are produced in us by powers within it.

D   Thus, nothing is found by analyzing a solid corpuscle,

corpuscle, but power or action: power, therefore, constitutes solidity, and is its essence.

The resistance or solidity of an atom, or of any mass, is twofold; it is a resistance to a nearer approach, and to a more distant recess. The parts of a hard mass resist being pressed together, or being pulled asunder.

This affords us only a two-fold modification, or application of the same idea.

Both these modifications of power are necessary to the existence of solidity, or a finite resisting extension, because either existing alone would be self-destroyed.

If in any active extension there existed *only* a power which directed all the parts within itself, the parts of the active extension would be carried inwards till the extension was lost.

If there were *only* a power which directed all the parts outwards, urging them to separate, the extension would spread, and the action lose itself in infinity.

Therefore the idea of a finite active extension, that is of any solid mass, implies the two powers,

ers, a power directing the parts inward or nearer together, and a power directing the same parts outwards, or further asunder, and these two powers equilibrating with each other, that neither prevails.

Thus, whether we analyze body by its sensible qualities, or by its component parts, we come at the same original principle, which is essential to it in every form, and presents itself at every view.

CHAP. III.

Error of the Doctrine of Impenetrability.

IMPENETRABILITY is that quality, aſcribed to matter, whereby every ſolid part excludes every other from within its own limits. This implies, that no ſolid part can be condenſed into a leſs ſpace than it occupies, nor extended ſo as to fill a greater, but that all have an equal and a neceſſary denſity, or equal quantities of matter within equal magnitudes; for if a body, from any ſize, can be rarefied into a greater, it can be re-condenſed into its former bulk, and this condenſation would be a penetration of its parts.

Solid parts being ſuppoſed impenetrable, the actual penetration of ſenſible bodies was attributed to the admiſſion of the penetrating ſubſtance into the pores with which matter abounds.

This doctrine affects the fundamentals of phyſics, and it is, therefore, of great importance to have its truth or error unequivocally ascertained.

D   Immaterial substances, it was allowed, could pervade bodies; but these were supposed incapable of acting on them by penetration or impulse, and to have influence only by volition. Material substances penetrating the interstices only, could not act but on the surfaces of the solid parts, and consequently their action could be supposed proportional to those surfaces only, and not to the mass.

E   But as the forces of moving bodies are proportional to their masses, the theory of any penetrating agent acting mechanically was precluded, because such could not penetrate the solid parts, by reason of their impermeability, nor, acting on their surfaces, could operate proportionally to the solid mass.

F   There are now, among those who in the main profess the Newtonian philosophy, some sceptics to this opinion, held essential in the earlier times of that system, and at present by its more rigid adherents.

G   This defection does not seem to have arisen from a desire to establish the theory of a permeation

meation of a moving caufe, againſt which impenetrability was fet up; but rather from a fecurity in the prefent doctrine of ACTION ON THE SURFACES, and of VIS INERTIÆ, imagining thefe to be now in themfelves fufficiently firm, impenetrability, their original fupport, is deferted, as having become unneceffary, and as being little defenfible. Thefe incautious architects trufting to the ftability the building has feemed to have acquired, from time and collateral props, allow the bafis to be removed, without being aware that they expofe the whole fabric to ruin; for when this barrier to a theory of permeation fhall be done away, THAT will be found to have fo many advantages over the notion of action at the furfaces, as will give it an indifputable claim to preference.

So little has been faid in defence of the doctrine of impenetrability, and fo fmall are its pretenfions to proof, that not many arguments can be required to invalidate its force.

In order to examine, and combat the idea of impenetrability, we need not infiſt upon power being the effence of body. If we merely confider matter as fomething that fills fpace, and by filling a fpace excludes other matter from the

F fame,

fame, this, without taking action into confideration, does not imply impenetrability.

K The terms impenetrability and folidity have, indeed, by fome, been ufed as nearly fynonimous; or at leaft it has been confidered that impenetrability was a confequence of folidity, and was implied in it; but this we fhall fhow to be unfounded.

L Fulnefs is an idea capable of intention and remiffion; the fame extenfion may be filled with different quantities of the filling fubftance; it may be more or lefs full, in all poffible degrees.

M Our ideas of the differences of denfities in bodies, is that of different fulneffes. Denfity is indeed, by philofophers, defined to be the number of folid parts within a given magnitude, each folid part being fuppofed equally denfe, and the differences in the degrees of denfity in the body to depend on the different quantity of pore.

N But affuming the equal denfity of folid parts, is begging the queftion for impenetrability: to prove the impenetrability of matter, it muft be *proved* that folid parts are of neceffity equally denfe, that is, contain the fame refiftence within the fame extenfion.

For

For if any two folid, imporous parts, of a given magnitude, can have different quantities of matter, whatever matter be, whether an inactive or an active repletion, then the given magnitude being filled with the lefs quantity, as in the part which contains lefs matter, can yet admit more or be more full, as in the part containing more matter; but more filling fubftance being admitted into an extenfion already full, is a penetration of that which at firft filled the extenfion. [O]

Now to determine this queftion, we may place it in two lights, each of which appears to me decifive. [P]

Firft, whether we can difcover any natural or neceffary relation which a given quantity of folid imporous matter has to a given magnitude. If a magnitude be given, what is the quantity of matter neceffary, without pore, to fill it, and why is that quantity alone capable of being contained in, and filling that magnitude. [Q]

We can, I believe, affign no reafon why among infinite quantities, one, more than another, fhould be neceffary to a given magnitude; why all poffible quantities, may not affume and fill that magnitude; or why any given quan- [R] tity of matter may not occupy and fill infinite different magnitudes.

s   It has been demonſtrated that a grain of ſand might be made to equal, in volumn, the orb of Saturn; or any given quantity of matter aſſume any given volumn, by forming ſucceſſive ſeries of porous parts. In this caſe, the magnitude of the ſolid body would not be increaſed, but its volumn only would be enlarged by the quantity of pore which interſected the ſolid continuity.

t   But if the above neceſſary limits (Q 35) cannot be aſſigned, there is no reaſon why a grain of ſand may not fill the orb of Saturn, or any greater magnitude, without any increaſe of its pores, but by the increaſed tenuity and expanſion of the ſolid matter.

u   We may next inquire whether we can conceive of a difference of denſity in ſolid parts, independent on pores. Of this there can be no doubt, when we conſider that children and the vulgar have no knowledge of inſenſible pores, yet they neceſſarily conceive of ſuch bodies as gold and ſteam, as differently denſe, or conceive of varieties of quantity in theſe different bodies, under the ſame magnitude.

Since

Since then we can conceive of no neceſſary limitation to denſity, or can diſcover nothing in nature which ſhould render it invariable ; and ſince we do familiarly conceive of different denſities in bodies, the doctrine of impenetrability muſt be conſidered as hypothetical, until it receives proof either from experiments, or from rational deductions.

Now experiments cannot poſſibly eſtabliſh this doctrine; for no difficulty, however great, experienced in penetrating bodies, can afford a concluſion, that greater forces than have yet been employed, might not overcome the reſiſtence which has only been inſuperable to an inſufficient power ; and it is not the actual impenetrability to any human force, or to any force which nature employs upon matter, which is intended, but a natural and neceſſary impenetrability, in itſelf unſurmountable by any poſſible force. Becauſe water has been found to have its parts reſiſt a very ſtrong preſſure, rather than mutually penetrate each other, no concluſion follows that they are not eaſily permeable to more ſubtil matters, unleſs the impermeability be conſidered as neceſſary and abſolute.

As to rational proof, I have never ſeen any attempted, nor can imagine any that can with
a ſhew

a shew of plausibility be offered: for the rational proof must consist in shewing that the penetrability of body would imply a contradiction; but the contrary of this appears, since we readily conceive of it as penetrable, and are unable to imagine how it can be deprived of its penetrability.

A   Conceiving of matter in its other, and true light, as an active fulness of extension, or as an extension full of action, the same method of reasoning will apply.

B   As we have no necessary bounds to our idea of density, so have we no limits to that of action.

C   Any given action, filling any extension, and producing resistence, and our idea of solidity, may be conceived as liable to be subjected to a greater action, and its resistence thus to be overcome and the solidity or extension penetrated.

D   We said no argument for impenetrability can be deduced from any actual resistence, because the action resisted, however great, is necessarily limited (Y 37).

E   But it will further confirm our objection to this principle, so far as it is supposed to rest on facts,

facts, to show that resistence does in fact take place in cases where impenetrability, and even solidity, are not supposed by any.

For if in any cases of resistence, solidity be found actually wanting, it will appear in all others unnecessary, and that which is certainly unnecessary and not certain in fact, we should conclude not to be.

The beginnings of all actions afford the facts to which we allude. Gravitation, muscular action, magnetism, all of these may be considered as resistences similar to that of hardness in bodies.

If with my hand I strongly compress two bodies together; if a heavy body be pressed to the earth, or a piece of iron to a magnet, is not the pressure, a hardness or solidity between the bodies, similar to, although it may be less than that which holds together two cohering parts of the same body? Yet no one traces the origin of these pressures to an impenetrability in muscular action, gravity, or magnetism.

Let us endeavour to bring together two like poles of a magnet, and we shall experience a resistence to their approximation. Does any one
say,

say, the extension between the magnets is impenetrable, or is solid, or is an inactive fulness?

K   Why may not, then, a piece of iron, which, between our fingers, resists their coming together, resist by a similar efficacy more strongly exerted, inasmuch as the effect is greater?

L   If magnetism were to act on our bodies, as upon iron, we should feel it; or were magnets endowed with sensation, they would feel that which resists their nearer approach.

M   The resisting extension between the two magnets is permeable to all the rays of light, and reflecting none, is unseen; but it is easy to conceive that the same power which resists the approach of the iron, might resist and reflect some rays of light. We should then have a visible object interposed between the two magnets, or extending round every magnet to the limits of its repelling sphere, as we have before supposed it might be a tangible one.

N   That which is tangible can be conceived as applied to our organs of tasting, of smelling, and of hearing, and as capable of exciting ideas of flavours, odours, and sounds.

Thus

Thus by a very eafy tranfition, by only fuppo- O
fing the magnetic action to be exerted on our
bodies, and upon light, inftead of being confined to iron, we fhould find it to poffefs all the
characteriftics of a folid body.

But would any change in the nature of the P
action deftroy the action; would the magnetic
virtue become an inactive folid by having its
effects extended to light and a human body?
This is not at all implicated.

Thus, we fee, an action in which no fuppo- Q
fition of folidity or impenetrability is involved,
may be conceived to affume all the qualities
of matter, by only fuppofing a familiar effect extended in its operation.

The author of the Preliminary Difcourfe to R
the French Encyclopedia reafons thus, in fupport of his argument to prove that fenfations are
the principle of all knowledge: " It is fufficient,"
fays he, " to prove that they are capable of being
" fo; for our fenfations are the moft certain
" facts, and facts, or acknowledged truths, are to
" be preferred before an hypothefis as the bafis
" of any deduction."

s   This juft and philofophical ground of argument is applicable to our prefent queftion.

t   Having fhewn that action is capable of producing the idea of folidity, and, by confequence, all ideas dependent on it, and which are only its various modifications, this is proof fufficient that action alone is the principle of all our ideas, and bafis of all our fenfations; becaufe action is actually found to exift in all, and any other origin is hypothetical.

u   The inactive folidity, the inert fulnefs of fpace, the unknown fomething, the myfterious fubftratum, about which philofophers have talked, being altogether inconceivable, altogether hypothetical, and wholly unneceffary; while action is conceivable, truly exifting and fufficient: if any one, without being able to prove thefe premifes erroneous, fhould yet infift on the folidity and impenetrability of matter in the fenfe here contended againft, he muft form his conclufion on grounds fo different from thofe which direct my judgment, that it would be fruitlefs for me and fuch a perfon to debate the queftion.

x   And he who denies a ftate of action in every thing we call folid extenfion, and choofes to infift on matter being an inactive fulnefs; or admitting
the

the ſtate of action, yet maintains the exiſtence of a ſolidity, which is inactive, and can ſupport theſe opinions by any reaſoning becoming a philoſopher, would open a door of knowledge of which I have, at preſent, no conception, and I ſhould be glad to receive inſtruction from him.

The doctrines of the IMPENETRABILITY of body, and of its INACTIVITY, ſerve mutually to ſuſtain each other, and it is difficult to determine which is prior, which ſubſequent. The notion of each primary atom being in itſelf inactive and cohering to others, ſo as to form larger bodies, by actions exerted at the ſurfaces only of the primary atoms, renders it neceſſary that thoſe primary parts ſhould not be liable to decompoſition, there being no action *within* them, by which any two parts could reunite, like that *without*, by which two parts coaleſce.

This opinion of inactivity within the primary corpuſcles, gave riſe to a fear, that if they ſhould be broken by any violence in the order of nature, or changed in wear, by having parts rubbed off into finer ſubſtances, and thus the original figure and ſize be deſtroyed, they could no longer ſerve to form the ſame kind of bodies which they did in their priſtine ſtate ; that, for example, water formed of old and damaged atoms, would be unlike

unlike the water at firſt produced by the new and perfect parts.

A   But this apprehenſion is founded upon the very extraordinary ſuppoſition, that although an action exiſts between the ſurfaces of the original corpuſcles, cementing them ſtrongly together, neverthuleſs, between any two parts of each original corpuſcle, which are yet more ſtrongly attached, there is no action at all.

B   But as we have ſhewn, and the thing is indeed evident, that action is no leſs eſſential to the exiſtence of the primary parts than to their union, we need no longer be ſolicitous to prevent their penetration or diviſion, ſince the ſame action, within, can preſerve the atom, although penetrated, from being deſtroyed, as without, can maintain or reſtore their coaleſcence which conſtitutes the forms of ſenſible bodies.

C H A P.

CHAP. IV.

Analysis of Motion.

SINCE all natural appearances confist either A of matter or motion, when, by a juft analyfis, we fhall have arrived at the fimple and conftituent parts of thefe, we fhall have got to the bafis of the fabric of the material world, to the foundation ftone of the building, beyond which, if we advance, we fhall pafs the line that feparates unconfcious from confcious being, and find ourfelves without the boundaries of corporeal nature.

To analyze motion is an attempt not very B confonant to the prejudices of the prefent times, which confider it as a fimple idea, therefore undiftinguifhable into parts, and undefinable; but that this is erroneous, will, I think, eafily be made appear.

We cannot conceive of any actual motion C without combining together thefe three ideas, a BEING which moves, a PLACE in which that being is, and the CHANGE of that place: hence the definition given. *(See Defin.)*

Let

D   Let us confider each of thefe three conftituent parts, and fee what each is in itfelf, whether each is not diftinct from the other two, and whether all of them are not necessary to the idea of motion.

E   Firft. A BEING which moves. A being may be confidered without motion: fuch is matter at reft.

F   Secondly. A PLACE. Place is that relation of one being to another, which confifts of *diftance* and *direction*. From a given body a mile north, in one place, two miles north, a fecond, a mile fouth, a third, two miles fouth, a fourth. Every variation of diftance or direction, produces a variation of place. *Diftance* and *direction* appear to me to be fimple ideas, and incapable of definition.

G   Thirdly. CHANGE. Change in any being is a ceffation of fome circumftance which was, and the production of one which was not. If a cubic figure be moulded into a fphere, the cube which was, has ceafed, the fphere which was not, is produced.

H   It is faid by fome, that change implies motion, and therefore cannot be a part of its definition,

being

being the very thing defined: to this I anfwer, we are fpeaking of the fenfible idea of motion, as it appears to our fight; now changes do appear to our view, and to all our fenfes, which give us no idea of motion. Changes in heat or coldnefs; in colour, flavor, fmell, found, hardnefs, foftnefs, pain, pleafure; in thefe, and many other ideas, changes do not produce ideas like that produced by a ball rolling, or a ftone falling. We may perhaps ultimately trace them to motion, but to infenfible motions; to motions which arife only in reflection, and conftitute no part of the actual idea of the change.

We can, therefore, conceive of change without conceiving at the fame time of motion. They are diftinct; as fenfible impreffions, however they may be derived from one caufe and have one effence.

Change is a generic idea, including many fpecies; motion, as a fenfible idea, is a fpecies of that genus. Motion is one fpecies of change, but there are alfo many others.

Change is therefore a neceffary part of the definition of motion; it marks the genus of the thing defined. Motion is a change; but as there are many fpecies of change, which of thofe fpecies

cies is motion? The anfwer is, it is a change of place. This marks the fpecies, and diftinguifhes it from change of colour, of temperament, of figure.

M All of thefe ideas (c 45) are diftinct from each other. The body is diftinct from the poffible places it may fucceffively occupy; the places a body may be conceived in fucceffion to occupy, are diftinct from the actual fucceffive prefence of the body in thofe places.

N All of thofe ideas (c 45) are effential to motion; otherwife which can you leave out of the idea without deftroying it?

O The combination of thefe conftitutes motion; otherwife what is that which is further neceffary? Thefe three, therefore, are the principles of motion refulting from its refolution.

P Since, therefore, change is an effentially conftituent part of motion, and by the definition change implies action, it follows THAT ALL MOTION IMPLIES ACTION, AND DEPENDS ON AN ACTIVE CAUSE.

Q Thus far we have confidered motion as a whole; we will next regard it in its parts. Every

motion

motion has a beginning, a middle, and an end. The beginning is a change from rest to motion; the middle is a continuance in motion; the end is a change from motion to rest. I am to shew that the beginning of motion is by an action began; the continuance of motion by an action continued; the end of motion by a cessation of the action.

The first position is admitted by every body.

That the CONTINUANCE of motion is by an action continued, will be proved if it shall be shewn that the continuance of a motion is nothing different from its beginning, in regard to any point of time assumed in the continued motion.

In what then it is asked does the beginning of motion consist? In a change from rest to motion? This is no definition, because it defines motion by motion; the beginning of motion does not consist in the rest which preceded it; therefore this definition amounts only to saying the beginning of motion consists in the beginning of motion.

I say then the beginning of motion consists in the beginning of change of place.

H   Now

x Now if any given portions of time and of space are affumed, a body beginning to move in the commencement of that time and in the firft portion of the space affumed, then and there begins that particular motion, and whether before the body began to move in that space, it was moving in other fpaces and times, has no relation to the motion in queftion; for this being in a time and space altogether diftinct, is a diftinct motion from any which might have preceded it immediately, as much as from a motion which preceded it a thoufand years before. It is therefore a new motion began; and fo it may be faid of every affumable point in the continued motion.

Y The term continued ferves only to connect any two diftinct motions, the end of one with the beginning of the other, but does not deftroy their diftinctnefs.

Z It is alleged on the contrary fide, *that a body in motion requires an active force to ftop it, but by reafon of its own inactivity not being able to exert any active force to ftop itfelf, it muft neceffarily continue in motion as a confequence of that inactivity, till fome other force ftops it.* This is I believe a fair ftatement of the argument ufed to prove that INERTIA and not ACTION is the caufe of continuance in motion; and in this way I have

have heard several eminent mathematicians argue, who confessed the *term* vis inertiæ was indefensible, retaining however all the incongruity in the *idea*.

If we only put together the premises and conclusion in this argument, who can believe one to result from the other? The premises state, that an active force is required to stop a moving body, and the conclusion is, that therefore a moving body is without any active force, and moves by its inactivity. Is it analagous to our reasoning in common affairs, to infer inactivity in the doing of that which it requires action to restrain? A

But we will examine the chain that leads to this conclusion, and point out where is its fallacy; and for greater precision we will put the argument in the syllogistic form. B

I.

Every body in motion requires an active force to stop it. C

No body in motion has any active force in itself.

Therefore, no body in motion can stop itself.

II.

II.

D A body in motion is unable to ſtop itſelf, by reaſon of its want of an active force.

But a body in motion of itſelf continues to move by reaſon of its inability to ſtop itſelf.

Therefore, a body in motion continues to move by reaſon of its want of an active force, or by its inactivity.

E The argument thus laid out, diſtinctly, in its parts, ſhews plainly where its error lies. The minor, or ſecond propoſition in the firſt ſyllogiſm, is directly an aſſumption of the thing in debate, a *petitio principii*.

F It aſſumes that A BODY IN MOTION IS INACTIVE, and thence it is proved, firſt, that it cannot ſtop itſelf, then that it continues to move as a conſequence of its inactivity; but that A BODY IN MOTION IS INACTIVE is the very thing in queſtion, and denied by the antagoniſt in the debate; THAT, therefore, is to be proved by ſome prior reaſoning, and not to be aſſumed in order to prove itſelf.

G Further, this ſecond propoſition is not only a *petitio principii*, but it appears no leſs evidently

to

to be contradictory to the major, or firſt propoſition. That a body in motion requires an active force to ſtop it, is the ſame thing with ſaying a body in motion has an active force in itſelf; it being plain, that whatever produces an effect, and requires an active force to deſtroy that effect, muſt be itſelf an active force; for the active force employed to *ſtop* a motion, would, in other circumſtances, *produce* a motion. Where then, it does not *produce* but only *ſtop* a motion, itſelf has deſtroyed the productive agency in the motion which it puts an end to, and its *own* producing power has been deſtroyed by the agency *it* deſtroys. This reſults from the truth evident in reaſon and in fact, that equal contraries, meeting, deſtroy each other.

The third propoſition or concluſion of the firſt ſyllogiſm is mere trifling, or ſomething worſe. It ſtates that a body in motion *cannot* ſtop itſelf; it having been before *ſhewn* that it requires an active force to ſtop it, and *aſſerted* that itſelf poſſeſſed none, it to be ſure follows that it cannot ſtop itſelf. But it has been ſhewn that a body requiring an active force to ſtop it, is an evidence that itſelf does poſſeſs an active force, and that if it cannot ſtop itſelf, it is not by reaſon of its want of a force to oppoſe to its own motion, but by reaſon of the force it poſ-

ſeſſes

fesses disposing it to move. Nothing can be more repugnant to reason and experience, than, in a moving body, to look for some power opposed to the motion it is found to have, and finding in it nothing opposed to its own motion, to ascribe its continued motion to the privation of any force which might oppose it, rather than to the presence of a force which preserves the motion. Do we ascribe a man's actions to the privation of a volition which might oppose them, or rather to the presence of a volition which directs them? Thus contrary is it to every other principle of reasoning, and to the plainest evidence, to assert a privation of force in a moving body which is capable of exerting force in every part of its continuity. This conclusion is trifling, inasmuch as it asserts *that a body in motion cannot stop itself*, because it is not *disposed* to do so, and it is erroneous, so far as it implies another cause, to wit, the want of an active force.

I   The second syllogism, which is built upon the first, must evidently fall with it.

K   The argument by which from the same premises (*i. e.* that a body in motion requires an active force to stop it) I draw a contrary conclusion, I shall next state in a similar manner, and subject every

every part to the fame critical review which has been taken of the other.

I.

Whatever requires an active force to ſtop L its motion, is difpofed to move.

Every body in motion requires an active force to ſtop its motion.

Therefore, every body in motion is difpofed to move.

II.

Whatever is difpofed to motion is poſſeſſed of M action.

But a body in motion is difpofed to continue in motion.

Therefore, a body in motion is poſſeſſed of action.

Thus it appears that the middle part of any mo- N tion (℺ 48) is action equally with its beginning.

The laſt part of motion is its TERMINATION. O It is admitted that all motion is terminated by an action contrary to the direction of the motion.

motion. It is admitted too that the moving body *acts* at the time its motion is destroyed. Thus the BEGINNING and the END of any uniform motion are confessed to be actions; but all the intermediate CONTINUATION which connects the beginning with the end is denied to be an action. What can be more unaccountable than that the same uniform motion should have action at its commencement, and action at its termination, yet be inactive during the whole time and space between the two extremes? How does the action at the beginning become changed into a privation of action? How does the privation suddenly become action at the end? Is it not more confonant to reason and analogy to ascribe to the whole continued motion, one uninterrupted action? Such a conclusion, true philosophy, we think, requires us to make.

When the reader considers that on this single question, *whether a continuance in an uniform motion in a right line* be the effect of INERTIA, or of ACTION, the whole of the theory of the philosophy at present received depends, he will not think I have entered too much into minutia in its discussion; and when he learns that if he admits, what to me appears self evident, and to arise from the definition of action, that motion can only be continued, and only subsist by action, he will then have

have it little in his power to refuse the remaining part of the doctrine laid down in the following pages; he will not think his attention misemployed in forming his decision, independent on the opinion of others, from the conviction of his own mind.

CHAP. V.

Action is an Attribute of an active Substance.

IN the foregoing chapters it appears that all phenomena are produced by action; we shall now proceed to show, that where there is action, there is some SUBSTANCE which acts. [A]

Our manner of conceiving things is analagous to the general division of phenomena into matter and motion. All our conceptions of being are referrible to SUBSTANCE, or ATTRIBUTE; to things which exist of themselves, or things which have being only in some other. [B]

To move or act, is an attribute; that which moves or acts is the substance. [C]

D   We cannot conceive an attribute to exift without a fubftance: for example, we cannot conceive of whitenefs as exifting without fomething elfe befides the whitenefs, and which we fignify by fome fuch mode of expreffion, as *the thing which is white, that to which whitenefs belongs.*

E   In like manner, we cannot conceive of action as exifting without fomething which acts, to which action belongs, or, which has the quality of action. Action is a verb, requiring its fubftantive, and this neceffity in grammar is founded on a prior neceffity in our manner of conceiving ideas.

F   We have traced all phenomena to action, as to a generic idea, comprehending under it all forms of matter and motion, as fpecies of that genus.

G   By this analyfis, that complex idea we have ufually denominated matter, and confidered as the fubftance, or fubftratum, to which motion appertained as an attribute, has been found to change its character, and to be itfelf an attribute.

H   The ACTION of a BODY IN MOTION is indeed the attribute of the body, and the body relatively

tively to its own motion, is truly a substance, having the attribute or quality of motion; but the body being a name signifying a combination of certain ideas, which ideas are found to arise from action (T 20), that action which is productive of those ideas whose combination we denominate body, is of the nature of an attribute; in other terms, body is to be considered as an attribute so long as it is considered as constituted of action.

To this attribute we must necessarily assign its substance. The actions which constitute body must be actions of something, or there must be something which acts.

WHAT then is this ACTIVE SOMETHING from whose agency we get the idea of body, or, whose actions constitute body.

Is it not sufficient that it is SOMETHING ACTIVE? When we say 'WHAT is this something,' let us ask ourselves what the question purports. Is it to get a name? that would be an easy task, we might give it an old name, or a new one, and define it to be something which acts to constitute body; but would the name render the idea more clear?

I 2   Do

M   Do we want a character, or description of the thing? Its character is that it acts; its description may be found in every sensation: it is colour to the eye, flavour to the palate, odour to the nose, sound to the ear, and feeling to the touch; for all our sensations are but so many ways in which this ACTIVE SOMETHING is manifested to us. No one sensible description, therefore, can serve for that to which all are common.

N   Do we ask what it is in itself, independent on being seen, felt, tasted, smelt, or heard? I reply, I know not; having no other means of perceiving it, but one of these, and in each of these it assumes its peculiar description.

O   At present we shall call that substance of which body is an attribute, the ACTIVE SUBSTANCE.

P   An INACTIVE substance, or substratum of matter, philosophers have imagined to exist, but have in vain sought to find. They saw the necessity of a substratum for the qualities of body, and having imbibed the notion of inactive matter, they sought for it an inactive substratum. At the same time they were persuaded that all the qualities of body were active POWERS they sought an INACTIVE support in which
they

they might inhere, a POWERLESS origin from which thofe POWERS might arife. Can it be wondered this refearch was fruitlefs?

They fought to conceive what muft of necef- Q fity be inconceivable; a fubftance, devoid of attribute, a being, without a quality, of which nothing could be declared that it DOES or IS. Having ftripped it of all conceivable conditions, they endeavoured to comprehend it. Might not the difappointment of fuch an attempt have been foretold? They fain, indeed, would have preferved, at leaft, the general idea of exiftence; but the idea of exiftence vanifhed when all its qualities were gone.

Our ACTIVE SUBSTANCE is the fubftratum R fo long fought for and with fo little fuccefs. We give this fubftance a quality by which it may be conceived. It ACTS. From this primary quality all the reft arife; ALL SENSIBLE QUALITIES ARE MODES OF ACTION, and action is common to them all. This renders the view of nature clear and confiftent; we have no longer to feek exiftence in non-entity, nor life among the dead.

One modification of ACTION produces MAT- S TER, another generates MOTION; thefe modifications

fications of action are modes of the active subftance, whofe prefence is action : matter and motion conftitute the whole of nature. THERE IS, THEREFORE, THROUGHOUT NATURE AN ACTIVE SUBSTANCE, THE CONSTITUENT ESSENCE OF MATTER AND IMMEDIATE NATURAL AGENT IN ALL EFFECTS.

Other inquiries may now occur, as whether this active fubftance be body or mind, material or immaterial, intelligent or uninformed; how it conftitutes matter, and how it produces motion; thefe things we fhall inveftigate in their places.

CHAP. VI.

On the Existence of an external, unintelligent Agent, with a Review of several Opinions, particularly that of BERKLEY, *on the Subject.*

WE shall, in this chapter, inquire whether A or not the primary substratum of matter or active substance possesses intelligence; and here it will be proper to notice what other opinions have been of late days maintained, concerning the actual external existence of the world, so as to give a comparative view of them with that offered in this work.

One opinion, very recently contended for is, B that material things exist without us, independent on being perceived, in the forms, and with the qualities which we ascribe to them from the impressions they make on our senses: a notion so unphilosophical and opposed to the first illuminations of science, which consist in correcting

the

the errors of judgment formed upon our senses alone, that it is sufficient to mention it, without undertaking its refutation in this work.

C   The doctrine maintained by Locke, is, that matter actually exists, solid, inactive, extended, figured and moveable, and these qualities are considered as primary and essential; but that it is in itself destitute of colour, flavour, odour, sound, and feeling, these, called secondary qualities, being said to exist only in our perception, and to be produced in us by the operation of the primary qualities.

D   Another doctrine published at Venice in the year 1763, by M. Boscovick, said to have been first thought of by Mr. Mitchell, and lately defended by Dr. Priestley, is, that matter is penetrable, and consists only of centres surrounded, at different distances, with spheres of attractive and repulsive powers, being unsolid and essentially active.

E   Berkley carried his abstraction still further, and argued that there was no intermediate agency between our minds and the supreme mind; that the world had no real being but in the reality of our ideas, and in the perception of the supreme mind which communicates them to us.

Of the first of these, which corresponds with F the Newtonian philosophy, it is not necessary to add any thing here, to what is contained in the preceding chapters.

The two latter agree that nothing INERT exists G in nature, but differ, in that one admits secondary active powers; while the other allows of no *mediate* instrument, but refers all our ideas *directly* to the volition of Deity.

The former of these two answers in general H to the opinion here maintained, and I shall briefly show, in this chapter, where appears to me to lie the error of the reasoning Berkley employs in defence of the latter.

Although the theory of Boscovick has a general I agreement with the doctrine of this work, inasmuch as it denies solidity, and asserts secondary powers without us, we do not accord in the manner of inferring or of treating of those powers; he confines his view to powers, and, not leading our mind to any substance as their support, gives us the obscure idea of an attribute without a substance; and in using the terms attraction and repulsion he falls into the error of substituting the names of effects for causes, and explains nothing; it not being more intelli-

K gible

gible how two diftant centres fhould act on each other, than how two diftant bodies fhould do fo.

K   It remains, that I examine the grounds on which Berkley denies all exiftence but mind, rejecting, not only fecondary qualities with Locke, and folidity with Bofcovick, but alfo extenfion, figure, action, and motion, as external beings, making all of thefe, ideas only, and wholly within a perceiving mind.

L   The limits pefcribed me do not allow of my quoting him at length, nor is this neceffary, as I fhall have to contend moftly with his principles, and if thefe be erroneous the conclufions muft neceffarily fall.

M   His fundamental pofition is, " THAT WE CAN PERCEIVE NOTHING IMMEDIATELY BUT OUR OWN IDEAS," and his argument may be conceived as continued in this chain.

N   " We cannot have any knowledge of external
" exiftence but by perceiving it immediately or
" mediately; external exiftences are not ideas,
" therefore not perceived immediately, therefore
" not known by immediate perception. We
" cannot have any knowledge of external things,
" mediately, by means of our fenfations, fince
" in

" in our fenfations we perceive nothing but the
" fenfations themfelves, and nothing without us
" which is not a fenfation, can have any refem-
" blance to fenfation, nor can be brought to
" our knowledge by means of that to which
" it has no refemblance; therefore we cannot
" perceive external exiftences immediately or
" mediately, therefore cannot know them."

He rejects the exiftence of primary qualities without us, alfo, on this ground, that folidity, extenfion, figure, and motion, have nothing in them uniform and permanent; the fame thing may be hard and foft to different organs; the fame motion fwift and flow; and the fame extenfion great and fmall, in different relations: for example, the fwift motion of a moufe would be a flow pace for a camel; the fize of a large dog would be very fmall in a horfe. The fame figure may be fquare, globular, and flat, to different views; hence he infers that thefe qualities cannot exift at all without us, fince they exift in no one particular manner.

He fays, THINGS do really exift without us; but by THINGS he means IDEAS, becaufe he knows of no other THINGS; and as exifting without our mind, they can exift only in another mind which

which perceives them, becaufe ideas cannot exift but as being perceived.

Q To prove the direct agency of a fuperior mind, and get rid of mediate caufes, he endeavours to prove that all our ideas are PASSIVE, that all our knowledge of ACTION is from our own volition; that, confequently, fince certain actions are performed on us without our volition, they are performed by another and fuperior volition without us.

R To the opinion of a fecondary agency, as an inftrument employed by the fupreme mind, he oppofes, firft, the want of any knowledge of fuch an inftrument, becaufe our ideas, which are all we know, are not inftruments of themfelves: next the unreafonablenefs of fuppofing God to ufe an inftrument, becaufe an inftrument, he fays, always implies weaknefs. We do not ufe a lever or pulley when we can eafily effect our purpofe by our hand alone.

S He takes pains to fhew that thefe notions are not repugnant to the common fentiments of men; but avers, that the philofophical notion of an unknown, inert fubftratum, as maintained by Locke, is fo (a).

_a See Dialogues between Hylas and Philonous.

I fhall

I shall first bring into question the principle т on which this ingenious writer builds his whole superstructure, 'THAT WE PERCEIVE NOTHING 'BUT OUR IDEAS.'

This position at first fight may appear evident, u because the expression to perceive ideas is familiar; but on examination it will be found to be, at least, inaccurate, and that inaccuracy, which the reasoner should have corrected, becomes a basis for erroneous deductions.

I mean, then, to show that ideas are not, as x this author takes for an axiom, the only objects of our perception, nor are they all, in strictness, OBJECTS of our PERCEPTION.

When we speak of an object of perception, y or a thing perceived, we distinguish between the object and the perception; between the thing, and the perceiving that thing. If we perceive ideas, ideas must exist, distinct from our perception. But our author, in another place, insists that ideas have no existence otherwise than as perceived; then, if their existence consists in being perceived, they are not distinct from the perception; but if not distinct from it they are the very perception itself.

We

z   We do not, then, perceive ideas or senfations, these are, the *very* perceptions, not their objects. To have an idea, is to perceive, but it is not to perceive the idea; the idea constitutes the perception. Since, therefore, we cannot be said to perceive any idea, we must, in every idea perceive its cause, and by the perception of the cause, acquire the idea.

A   Although the cause is distinct from the sensation it produces, we do not perceive the cause as distinct, we perceive it only in and by the sensation, and in the sensation only do we discover and know the cause; hence we conclude, that since the objects of our perceptions are not ideas, that is, are not the perceptions themselves, but their external causes, we do, immediately, perceive, and know external objects.

B   These objects are, resistence, extension, figure, and motion, perceived by sense, but themselves not sensations. As we are capable of moving our own limbs by volition, we discover resistence when we *will* the motion and it does not take place; as, if a man's hands are bound together, he finds the resistence of the cords by his inability to separate his hands, and this resistence is distinct from the sensation of hardness : it is not a sensation, but the negation of an accustomed sensation, fenfation, which the motion would have excited. Refiftences have certain limits, we do not find them every where; a poft refifts me in one direction, but by a little deviation I move freely. The variety of thefe limits, and their relations, are extenfions; from extenfion of neceffity refults figure. Thus refiftence is difcovered by motion impeded, and refiftence being limited and various, implies extenfion and figure. Now motion is without us; for fince any particular idea is not always within us, its caufe is not always within us; in producing the idea the caufe from without muft have entered within; the caufe muft, therefore, have moved without us, from where it was, in order to arrive within us, where it was not. Motion is therefore really without, if by without we mean any thing; therefore the refiftence, extenfion, and figure, are alfo without us.

To the objection that all thefe things are variable, and having no particular and fixed being, have no being (p 67), fince, for example, the fame motion may, in our *ideas*, be both fwift and flow, but cannot be *actually* both, there is really no motion; I reply, this makes as much againft motion exifting in idea, as externally; for if a motion can, as the objection fuppofes, be both fwift and flow in idea, it can be fo independent

pendent on our idea, since whatever can be conceived, is possible to be. Relations are permanent only to the same relative objects, and variable to varying objects. A motion may be, at once, both swift and slow, compared to two other motions, but compared with the same motion it is permanently the same. If we take the relation in which a man may stand: to John, he is a father, to Peter, a son, to William, a master, to James, a servant, to George, a subject, to Frederic, a king. Has this man no existence, because in our ideas he is father and son, king and subject? And is there no father, son, master, servant, king, or subject, because they are all found in one man? So we must conclude on the premises which deny motion because the same motion is at once swift and slow.

D   The same answer will apply to the same objection, concerning resistence, figure, and extension; all of these are necessarily variable in relation to different objects, and permanent to the same objects. Any individual thing has indefinite different magnitudes relative to indefinite other different magnitudes; every one of these is real, that is a real relation to its proper object.

E   Since, therefore, things which are not ideas
do

do exift without us (p 67), to wit, the actions which produce ideas, there are external exiftences independent on being perceived by us.

The pofition that all our ideas are paffive F (Q 68) is contrary to felf-evidence. Are not all our ideas changes taking place in our confcioufnefs? And can change exift without action?

We are, it is true, paffive fubjects of the G action in which ideas confift; but to call the ideas paffive is to talk of a paffive change, virtue, or power. Let us illuftrate this by an example taken from our author.

He fays, " When I ftir my finger it remains H " paffive; but my will, which produces the " motion, is active." How does it appear that the will is active? By producing motion, fays our author; then whatever produces motion is active? And does not the finger produce motion as well as the will? Does it not move the air or other bodies which the will cannot move, as it can the finger? Is there not the fame evidence of action in the finger as in the will? If the finger is paffive to the will which moves it, is it not alfo active to thofe things which it moves?

I. But

I   But it was neceſſary to prove MOTION not to be ACTION, in order to exclude ſecondary cauſes, and eſtabliſh the notion of the direct agency of Deity; Berkley therefore argues in this chain : " Firſt, " Are we not paſſive in all our ſenſations ? Can " we avoid ſeeing when we open our eyes, or " feeling when an object touches us ? Secondly, " Are not therefore all our ideas perfectly paſ-" ſive and inert, including nothing of action in " them ? Thirdly, And are ſenſible qualities " any thing elſe but ideas ? Fourthly, And is " not motion a ſenſible quality ? Fifthly, Con-" ſequently no action."

K   To each of theſe inquiries our author makes his antagoniſt aſſent, and the concluſion is undeniable ; but we cannot extend our complaiſance ſo far. Admitting the firſt poſition, the ſecond is no conſequence ; we are paſſive in receiving our ideas, our ideas are not therefore paſſive ; paſſion implies action, and the idea is the action, of which we are the paſſive ſubjects. The marble is paſſive in the hands of the ſtatuary, but it is by means of actions it aſſumes various figures. Thus the very firſt ſtep from the premiſes is erroneous. The next " Are ſenſible qualities " any thing elſe but ideas." We have ſhewn that *ſome* ſenſible qualities are external objects of ſenſe, not ideas, and among theſe, motion is

one :

one: thus his second step is also denied. Ideas are not passions, but actions; but if they be passions, motion is something more than an idea, it is proved to be action; consequently, the conclusion that motion is no action is proved to be erroneous.

Since, therefore, motion is action, and motion L does not depend always on our will, we are acquainted with other actions besides those of our volition, (Q 68).

Hence is destroyed the ground on which our M author inferred a direct superior volition without us, in all those actions which our volition did not perform.

He says, " If motion be no action, can you N " conceive of any action besides volition ?" But motion is action, therefore the premises are done away.

The objections (R 68) against a mediate in- O strument, or secondary cause, are already, in great part removed, since we know, in fact, of such an instrument.

If, by my volition, I produce a motion, that P motion is not a volition, but it produces another

motion

motion, and is, therefore, an intermediate agent, an involuntary and fecondary caufe.

Q The other objection, ' that it is unworthy of ' Deity to fuppofe him employing an inftrument,' takes a ground on which human reafon is very incompetent, and it will have no weight, if it can be proved that, in fact, the Deity does employ inftruments.

R I afk, therefore, is not the human mind created to fome end? And is it not an inftrument to the end for which it was made? God therefore employs inftruments to his ends. Why not, then, an inftrument to act on mind, as well as mind the inftrument to fome other end. God can produce ideas in mind without fecondary caufes. Could he not have brought about the end to which mind was created without this means? The inftrument here fhows will, not weaknefs. Why not an inftrumental univerfe as well as inftrumental minds? And how, from our ufe and need of inftruments can we prefume to form a pofitive inference to the Deity? Can we trace the connecting chain?

S Berkley admits an external caufe of fenfation, but contends, that that caufe is Deity alone: we imagine there is a mediate power, unintelligent,

gent, directly operating upon our minds, and producing according to regular laws, changes in the mind, of which changes it is conscious, and which are its sensations, and one source of its ideas (b).

We say then, that, although we find in each, and every individual sensation, a power producing and an action exerted, yet that agency is not such as leads us to infer the direct operation of intelligence. We conceive of intelligence, only by attention to our own minds, and there is nothing in any colour, taste, sound, feeling, or motion, which implies any thing analagous to our mental operations, or to volition. I say there is nothing in these sensations individually, which implies the direct presence of a thinking mind.

Therefore, in these sensations not discovering the traces, not seeing the characters of intelligence, but finding only action present and necessary, our inferences go no further than our observations warrant us to do, and we conclude in all these things an action only, and that action unintelligent.

b A sensation may be defined, a consciousness accompanying an external impression; an idea is a present consciousness of a past impression, external or internal.

We

x   We do not confider whether God could, without any inftrument, have produced all our ideas, and given to us that perception which we have of an eftablifhed order of nature: without doubt we ought to admit he could.

y   But this poffibility admitted affords no juft grounds for the conclufion that he has done fo, againft the ftronger evidence of reafon and facts that he has not.

z   What is it, then, from which we can infer intelligence? We know intelligence intuitively, by our own confcioufnefs; we infer it, without us, by perceiving effects, fuch as we find our own intelligence produces. IT IS BY AN ORDER OF SUCCESSIVE EVENTS LEADING TO AN END, that we infer intelligence, becaufe fuch is the operation of our own intellect.

A   Although we difcover no intelligence in any one fenfation, and can infer none in the agent which directly produced it, yet in the order and concatenation of events we gather this idea, and by this order it is that our minds are led through the intermediate, fubordinate, and unconfcious agency, manifeft in individual appearances, to a directive, governing, fupreme intelligence.

Nothing

Nothing in this idea shocks our reason, or offends our moſt exalted ſentiments of Deity. We are not, indeed, to make our ideas of dignity any ground of inference to Deity; but neither can we admit of ſuch ideas as to ourſelves appear low, and unworthy; and it ſeems to me, that a ſubordinate agent, conſtructed as the matter of creation, inveſted with perpetual laws, and producing agreeable to thoſe laws all the forms of being, through the varieties of which inferior intelligences can, by progreſſive ſteps, arrive ultimately at the ſupreme contriver, is more agreeable to our ideas of dignity, and tends to impreſs us with more exalted ſentiments, than viewing the Deity directly, in all the individual impreſſions we receive, divided, in the infinity of particular events, and unawful, by his continual preſence in operations, to our view inſignificant and mean.

There is one more argument in proof of a mediate agency, which, I think, alone concluſive, as it ſhows the contrary, if admitted, leads to conſequences which, aſſuredly, no one will defend.

Whoever denies the inſtrumentality of a mediate cauſe, and aſcribes all his ſenſations to the direct volition of the ſupreme intelligence, muſt,
conſiſtently

confiftently with his principles, deny the exiftence not only of body, but of all other minds befides his own, and that from which he receives his ideas. If he denies that a material world exifts without him, a globe of earth, ftars, planets, mountains, trees, houfes, and the bodies of animals, he muft of neceffity deny the powers with which thefe corporeal forms are endowed. If no fun, nor earth, no gravity to preferve the fyftem; if no vegetable world, no principle of vegetable life to govern it; if no material animal forms, no fpirits to prefide over them; if no bodies of men, no fouls to inform them, he muft be himfelf the only object in creation, a folitary creature, to whom the fupreme mind has thought fit to impart a certain order of ideas from his own effence, prefenting to him the glorious picture of the univerfe.

The exiftence of any other mind like his own muft to fuch a man be as unknown as are the inhabitants of the Georgium Sidus; he being altogether excluded from any commerce with them, and they with him. Since to every mind God immediately communicates from himfelf all its ideas, all intercourfe with each other is cut off, every one is an ifolated being in creation, and muft conclude himfelf the only favoured communicant of Deity.

For

For if he concludes all the appearances he sees of human bodies, performing certain motions, are only ideas WITHIN him, how can he assign to each of those bodily forms a subordinate mind like his own, existing WITHOUT him? If the bodies are within him an idea only, the mind appertaining to those bodies must be within those ideas which are within his mind. But if he allows of external subordinate minds, how does he infer them? How from sensible appearances, which are all within him, can he be led to beings without him, which are insensible? By what medium does he acquire a knowledge of these external minds?

His own body is, with him, merely an internal idea; other bodies the same. He is conscious of volition and agency, by which certain ideas are, in a degree, within his own power; he is subject to other impressions or ideas not in his power to control; he thence infers a superior volition to his; but these ideas in which he is passive are within; the volition, therefore, which directs them must operate within him, and cannot lead him to any knowledge of minds without.

Lastly, the premises on which our author reasons against the existence of a secondary cause

cause without us, go, if admitted, to the rejection equally of a primary. He rejects the existence of external matter, because we can have no knowledge of it, all our knowledge being confined to our internal ideas; for the same reason we can have no knowledge of an external Deity, nor does the way in which he reasons to the knowledge of a Deity do away the charge of inconsistency with his principles. 'We perceive,' he says, 'our volition to be action, and we perceive actions which are independent on our volition; but as we do not know any agent but volition, those actions which *we* do not will arise from an *external* will.' But we do not perceive or feel that external will, as we do our own volition; we cannot, as it is without, know it by our ideas within. The actions from which he infers an external volition, are they not ideas within him? And if they arise within him, independent on his will, he cannot see their derivation or origin; they are to him original ideas. He cannot trace them to a will without him, of which he can have no knowledge; there is no chain from within to without; all his knowledge exists, arises, and ends within himself. If he says he infers a will without him because the ideas independent on his own will *must* have an origin in some will, how does he know that those ideas which he does not perceive to arise

from

from volition *muſt* arife from volition? This is mere hypothefis, afferting for knowledge what he does not *perceive*, and therefore cannot know. Without an external *paſſive* exiftence no proof can arife of external *aƈtion*, and only through a world without us, can we trace the all-governing mind. It is the magnitude of creation which gives us correfponding impreffions of the greatnefs of its author, and by contraƈting all things into ideas within a human mind, we contraƈt our conceptions of Deity into the fame narrow point of view.

Thus I have endeavoured to remove fuch objeƈtions from pre-conceived opinions, as might lay againſt the belief of an aƈtive, unintelligent fubſtance, the exiftence of which I have already proved, and ſhall proceed to confirm by new evidences.

CHAP. VII.

The active Substance an immaterial Existence, not Mind, but intermediate, and related to Matter and Mind.

A ALL existence must be either material or immaterial, since immaterial implies only the negation of those qualities which constitute matter, and comprehends all existence in which those qualities are not found, and which, therefore, is not matter.

B But every being is not necessarily either matter or mind, because each of these terms implying positive characters, whatever substance may be conceived to want those positive characters, and to possess others, different from either, will belong to neither of these heads, but will be a distinct order of existence.

C Matter is a being, as a whole quiescent and inactive, but constituted of active parts, which resist separation, or cohere, giving what is usually denominated solidity to the mass.

Mind is a fubftance which thinks.

A being which fhould anfwer to neither of thefe definitions would be neither matter nor mind, but an immaterial, and, if I may fo fay, an *immental* fubftance.

Such is the active fubftance which we have difcovered by the above analyfis of matter and motion; we have already concluded it to be un-intelligent (v 77), becaufe we have no grounds from whence to infer that it thinks; it therefore is not mind, neither can it be matter; for being that which refults from the analyfis of matter, it would be abfurd to fay, that matter was by analyfis refolved into matter, fince that is no analyfis which difcovers nothing beyond the thing propofed to be analyzed. Cohefion is an effect, in which we difcover an active caufe, and thence infer an active fubftance; but this active fubftance cannot be a cohering fubftance, for then the cohefion would ftill be an effect of which we fhould have to feek the caufe, and we fhould have made no advance towards difcovering the caufe of cohefion. The elementary principles of things muft be different from their compounds, and muft, themfelves, be without thofe properties which the compounds derive from the union of the principles; the active

fubftance

substance therefore is devoid of cohesion, or, what is called, solidity; wanting this, it wants that sort of action among its parts which constitutes cohesion; but since it is an active substance, and has not those actions of its parts which matter possesses, producing inertia of the whole, it must have some other action, to wit, an action as a whole quantity. Thus it is without cohesion, which matter possesses; and is possessed of activity as a whole, of which matter is destitute; it wants therefore the positive characters of matter, and possesses another character which matter does not possess (B 84), it therefore is not material.

G   This substance, neither mind nor matter, is, in some respects, related to both, and may be considered as an intermediate existence; it is related to matter inasmuch as both the active substance and matter are active, although in different manners; the activity of the one being the result of the activity of the other, as a compound results from the union of its elements. It is related to mind also, as both are in certain manners active, and as both have the negative quality of immateriality.

H   If it be alleged that such a substance is a new order of being, I reply, as far as I know, it is;
but

but I do not fee in this, any argument againſt its exiſtence. We do not, here, propoſe to record ancient opinions, or deliver eſtabliſhed ſyſtems, but ſeek, by the exerciſe of reaſon, to improve the ſtate of ſcience; the doing of which neceſſarily implies both the renunciation of former errors, and the diſcovery of new truths.

We might, in the extenſive ſenſe in which the ancients applied the term mind, call the ACTIVE SUBSTANCE the MIND of the univerſe, and in every particular thing, the mind of that thing; the active principle by which a mountain coheres and gravitates, by which a bullet flies, the earth moves, and a tree grows, would be the mind of the mountain, the bullet, the earth, and the tree; for they, in general, termed all active principles mind: but this ſenſe of the word is not in common adopted, in the preſent day; and it is both more ſafe to adhere to general uſage, as far as poſſible, and important to diſtinguiſh by different appellations, things ſo different as thinking and unthinking ſubſtances.

PART

PART II.

Of Action, and the Manner in which the ACTIVE SUBSTANCE *produces Matter and Motion, investigated by Inferences from Effects to Causes.*

INTRODUCTION.

THE analysis attempted in the first part is to be considered less as a work of reasoning than of observation. It is a decomposition of sensible impressions, unravelling the intricacies of our compound ideas, and bringing them into a simple point of view. We laid down no data from which to reason, but merely examined and developed by a kind of mental experiment, the various ideas received through our external organs. It is intended, in great part, to show a very simple and obvious fact, that matter and motion

motion are active; but this evident fact being denied in a favourite fyſtem, oppoſed by ſtrong prejudices, obſcured by an ingenious converſion of terms from their genuine import, and the truth born down with the weight of an authority to which the age yields an implicit faith; theſe things rendered it expedient to inveſtigate minutely, and turn on every ſide, a poſition, which, but for them, would have carried its own evidence.

B   We conſider it now as a fact eſtabliſhed, that every thing in nature is active, and the inference from hence that all things exiſt, and all changes are produced by an ACTIVE SUBSTANCE, is alſo aſſumed as a datum. Upon theſe premiſes we ſhall now proceed to inveſtigate further the nature and laws of the ACTIVE SUBSTANCE, in producing the phenomena of the world, by drawing evident and undeniable inferences from effects to their cauſes.

C   When, however, we ſpeak of ACTION as univerſal, it is always to be underſtood that there are certain ſenſes in which every thing may be ſaid to be INACTIVE. Every thing is active, but not active in every manner; and a right knowledge of theſe diſtinctions is eſſential to juſt conceptions of nature.

C H A P.

CHAP. I.

On the Rules of Reasoning from Effects to their Cause.

THE knowledge of the connections that link A events together, and constitute a continued chain in nature, is the sum of philosophy.

A CAUSE may be defined an event which B necessarily precedes another; an EFFECT is that other event which is necessarily preceded by the cause, and necessarily follows it.

There is no small obscurity pervades almost C the whole of physics in respect to the division and doctrine of causes; and to this obscurity the imperfect and undecided state of theoretical knowledge is owing. As it is our purpose to endeavour to advance forward in science, we propose to lay down such rules as may conduce something to this end, rather than follow those heretofore established.

D   The rules of reasoning in philosophy, as laid down by Sir Isaac Newton, and generally adopted, are as follow:

First. *To admit of no more causes of natural things than are both true and sufficient to explain their appearances.*

Secondly. *Therefore to the same natural effects, the same causes must, as far as possible, be assigned.*

Thirdly. *Qualities which admit of no change in degree, and are found in all bodies as far as we can observe, are to be esteemed universal.*

Fourthly. *Propositions collected by induction from phenomena are to be admitted before any contrary hypothesis.*

E   These rules are, no doubt, just and useful, as far as they go, but they appear to me to be imperfect; they are calculated, indeed, for that foundation of philosophy on which Sir Isaac proposed to rest his inquiries, *experiments;* but are inadequate to the more extended pursuits of rational investigation. The first rule requires that we should admit no more causes of natural things than are both true, and sufficient; but by what *rule* are we to determine concerning the

the TRUTH and SUFFICIENCY of caufes? Surely a rule to conduct ourfelves in thefe points is the moft important of all; for it is to no end that we confent to admit of that only which is true and fufficient, if we have no ftandard whereby to determine what are to be confidered as having thefe requifites. Here is therefore a fundamental deficiency in thefe rules, a deficiency which expofes us to every fort of delufion and error; they only direct us to follow a fingle path rather than many, and command that this path fhould be the right one, but give us no marks whereby we may diftinguifh the right path from many others, which are wrong.

It is probable, indeed, that in a philofophy propofed to be built on experiments alone, experiments would be confidered as the only teft of the truth and fufficiency of caufes; but this precludes all inveftigation purely rational, and can therefore avail us nothing in fuch an attempt. At the fame time, if experiment be confidered as the only teft of the truth and fufficiency of caufes, it muft be obferved, that our author in forming his firft principles deviated from his own rules; for it was not poffible to eftablifh thofe principles by experiment; and they are purely refults of a mental procefs, I will not fay of reafoning, becaufe I think them repugnant to reafon,

[94]

fon, and regard them as hypotheses fabricated in the mind to explain appearances. I speak here of the VIS INERTIÆ, or INERTIA of matter, a principle incomprehensible in itself, and wholly incapable of an appeal to experiment; and the VIS IMPRESSA, or principle of action, which, as defined and explained, is equally obscure and unfounded in fact. This incongruity, into which Sir Isaac has fallen, cannot be too much insisted upon, because it affords a subterfuge for error, and perplexes the minds of inquirers, who continually recur to the plea, that the Newtonian philosophy is founded on experiments, and that its principles are only the assertion of indubitable facts, when the truth is the very contrary. Sir Isaac does indeed in so many words disclaim and renounce hypothesis, and professes to be guided only by facts; but his first principles, even if true, cannot be pretended to be facts, they must be rational truths, defensible by reason as deductions from facts: but to this kind of proof, we have, in another work, shewn they can lay no claim.

G What appears to be yet wanting in physics, is some test more certain, and more general, than has as yet been given of the truth and sufficiency of causes; and to attempt to supply this desideratum,

let

let us examine further, what is the nature of the relation of cause and effect.

The necessary connection between two events, one being prior, and the other subsequent, constitutes the former a cause, and the latter an effect.

Under this general character several kinds are included, which may, for the most part, be comprehended in the four following, rational, experimental, remote, and proximate.

The connection between cause and effect may be *perceived* by the mind: thus the solution of sugar in water is a cause of sweetness in the water; the connection is traced in the contemplation of the divided parts of the sugar diffused through the water, and imbuing it with its own quality. Or the connection may be *unperceived*, as in the change of saccharine substances into ardent spirits, in the vinous fermentation. When the connection is unperceived, the mind *infers* the connection *a posteriori* from the fact observed; we conclude cold to be the cause of the change of water into ice, not because we perceive the connection between cold and fixedness, but because we perceive in fact the two events, a certain degree of cold,

and

and the production of ice, do always happen in the order of prior and subsequent.

L   This distinction gives us two sorts of CAUSES, which we may denominate RATIONAL and EXPERIMENTAL. The rational, being those whose connection with the effect is perceived, the experimental, those where it is only inferred from the fact first experimented.

M   Of these two it may be observed, that the former affords us grounds for reasoning *a priori*, and for rational proofs, in cases where the opportunity of making experiments may be wanting. Knowing water to be a substance devoid of colour, taste, and medicinal qualities, we can certainly infer, *a priori* any colour, taste, or medicinal quality to be imparted to the water, by a solution, in it, of known substances. The latter admits of no reasoning apart from experiments; we cannot draw a single inference without hazard of being wrong, but must, at every step, make trial of the fact, to assure the truth of our proceedings. We cannot from the sight, smell, or taste, assure ourselves of the nutritive, or deleterious qualities of any unknown plant, but must make the experiment before we can determine.

Another diſtinction between cauſes and effects is into REMOTE and PROXIMATE. In a chain, each link is connected immediately to the link neareſt to it, and remotely to others at a greater diſtance. Events ſucceed to each other in nature in a certain ſeries; each event is connected immediately to that which directly precedes, and directly follows; and remotely to others, ſeparated from it by a greater or leſs number of intervening events. The remote cauſe of an eclipſe of the ſun is the force by which the moon performs certain motions relative to the ſun and earth; the proximate cauſe is the interpoſition of the moon's body in a line between the ſun and earth.

There are other varieties among cauſes, which I ſhall not minutely inveſtigate. Several links may be connected remotely or immediately with one; ſeveral cauſes remote or proximate may concur to one effect, and concur in different degrees, and different manners. Hence, cauſes may be diſtinguiſhed, as they are predifpoſing, occaſional, inciting, partial, primary, ſecondary, efficient, mediate, and an abundant diverſity of others, which accurate obſervations would lead to diſcriminate.

P   I have only to consider the diſtinction of cauſes as they are experimental or rational; and to one or the other of theſe heads all caufes belong. Experiment is the only ſtandard of experimental cauſes; when we fee an event uniformly followed by another, we are ſatisfied the former is a cauſe of the latter, both TRUE and SUFFICIENT, although we do not perceive how the two events are connected, nor in what confiſts the ſufficiency of the caufe; we *infer* that the connection does exiſt, and the fact proves the caufe to be a ſufficient one. But when we *reaſon* to the diſcovery of a caufe, and either decline or are unable to employ experiments, we muſt have a different ſtandard of what is true and ſufficient.

Q   This ſtandard can be no other than the *perception* of the neceſſary connection between events; in other terms the *perception* of ſomething in the caufe, which involves the effect; for fince, in this cafe, all experiments from which the CONNECTION between events can be *inferred*, are precluded, that CONNECTION can no otherwiſe be known, than by being perceived as an intuitive truth.

R   Hence, when the caufe of any effect is fought, if we can, by experiment, aſcertain what appearance,

pearance, or event, always, and directly, precedes the effect, we shall have found an experimental cause. If we can *perceive* what it is which necessarily and alone involves that effect, we shall have discovered its rational cause: for example, impulse is an experimental cause of motion, but not a rational one, since we do not *perceive* how the impulse of one body involves the motion of another; but if besides the impulse, as it appears to sense, we can find an idea which does necessarily, and alone, involve the motion of the impelled body, this idea will be the rational cause of the motion in this body; it will be TRUE, because the idea perceived being the only one which involves the effect, *that*, and no other must be the cause; it will be SUFFICIENT, because it does actually involve the effect.

If one, of two distinct substances, possesses a quality, of which the other is wholly a privation, the union of these two substances necessarily implies, in the whole, formed by the union, the presence of that positive quality, which, before the junction, belonged only to one substance. Thus whiteness being a combination of all the primitive colours, is considered as a privation of each single colour; the union of any primitive colour therewith imparts to the whole, thus united,

united, the pofitive colour, red, orange, blue, or green, which has been incorporated with the white. Water is a privation of fweet, bitter, acid, or of any medicinal quality; a fubftance poffeffing any of thefe qualities, being diffolved in the water, the new mafs poffeffes the actual qualities, which the diffolved fubftance has fuper- added to the privation of them, in the water.

T   And, in general, from the combination of a fubftance, having a pofitive quality, with one which is a privation of that quality, we infer the prefence of that pofitive quality, in the whole compound. Thus we argue from the caufe, the incorporation of the two fubftances, to the effect, the prefence in the whole compound, of the quality, of which, before this union, one fub- ftance only was poffeffed, and one deftitute.

U   If a fubftance, in its own nature devoid of a certain quality, be found accidentally poffeffed of that quality, we neceffarily infer that the new quality has been fuperadded, and the fuperad- dition of the new quality implies, alfo, the pre- fence of the fubftance to which that quality belongs, fince no quality can exift apart from its proper fubftance.

X   Thus, if a mafs which before was white,
becomes

becomes red or blue, we infer that some substance, red, or blue, has been mingled with the white. If water, before infipid, and unmedicinal, acquires taste or medicinal qualities, we infer the addition of some substance from which the quality has been imparted.

The same necessary connection with the former y (T 100) is here traced in an inverted course. From the presence of a new, and accidental quality, we infer the presence of the substance in which that quality inheres. Here we argue from the effect, the new quality apparent, to the cause, the superaddition of that quality, by the junction of a foreign substance, to which it belongs.

We can have no other ground of inference z than this, the change of some quality, either in figure or place, and assuming a new form, or being united to a new substance. We cannot conceive of any thing, as bitterness, pain, or motion, being produced from nothing, nor can we conceive of one thing, as found, being produced from a different one, as odour, or of motion, as proceeding from rest. An effect, which is only a new appearance, of something *where* it was not before, or in a *manner* in which it did not before appear, can only be comprehended by supposing the thing did before exist,

and

and has, in order to the production of the effect, arrived into a new situation, or has assumed a new modification, or figure. Probably we shall never comprehend how motion produces senfation, because in motion we perceive nothing that can, by any change of place or figure, involve senfation.

From what has been said, we may lay down the following rules in the investigation of rational causes, from their effects.

RULE I.

A A RATIONAL CAUSE MUST CONTAIN IN IT THE NATURE AND QUANTITY OF THE EFFECT.

For otherwise, in the effect, a new nature or a new quantity, must be produced from nothing, which is what we cannot comprehend, and therefore cannot admit.

RULE II.

B A RATIONAL CAUSE MUST BE PRESENT IN TIME AND PLACE WITH ITS EFFECT.

RULE III.

C WE ARE TO ADMIT OF BUT ONE RATIONAL CAUSE.

For

For when reason is satisfied with one, to what end should we seek for more?

RULE IV.

EFFECTS DIFFERENT ONLY IN CIRCUMSTANCE, BUT THE SAME IN NATURE, ARE TO BE ASCRIBED TO THE SAME CAUSE DIFFERENTLY CIRCUMSTANCED, AND NOT TO DIFFERENT CAUSES.

This is a consequence of the third rule.

RULE V.

EQUAL CONTRARIES MEETING DESTROY EACH OTHER.

Contraries are things which cannot exist in union, whence the proposition follows.

The first and second rules are contained in the positive principle that we must perceive the necessary connection between a cause, admitted as rational, and its effect; for if the cause and effect differ in nature, or in quantity, in time, or in place, there is between them a chasm through which the mind can trace no connection. The third and fourth rules belong to the negative principle, admitted in philosophy, that we are not to seek superfluous causes, or to suppose nature has done any thing in vain. The fifth rule is evident.

CHAP.

CHAP. II.

In what Manner the ACTIVE SUBSTANCE *is active, and how it acts.*

A   WE are to diſtinguiſh, here, between the activity and action of the ACTIVE SUBSTANCE, and the activity and action of bodies; the one being primary and eſſential, the other derivative and ſuperadded.

B   We are alſo to diſtinguiſh between *activity* and *action: to be active,* is to poſſeſs the quality of activity, or to be capable of acting; and this may belong to *one* ſingle ſubſtance.

C   *To act,* is to exerciſe the quality of activity; and to exerciſe this quality is to communicate it; for whenever we ſay a thing acts, we mean that it induces an action on ſome other thing; therefore *two* things are neceſſary to action, an active thing, communicative, and a paſſive, recipient.

We may, therefore, confider firft, in what D conſiſts the *activity* of the ACTIVE SUBSTANCE in itſelf; and afterwards how it *acts* or communicates its own quality to another.

Every thing muſt be, in its own nature, either E diſpoſed to reſt or motion; conſequently the ACTIVE SUBSTANCE muſt be confidered as being, naturally, either quiefcent or motive.

But it cannot be naturally quiefcent, or diſpoſed to reſt; for then it could not be active, becauſe activity, which is a tendency to motion, cannot originate in a tendency to reſt.

Therefore the ACTIVE SUBSTANCE is by G nature motive, that is, tending to motion.

We learn, from obſervation of bodies, two H conditions in which they have a tendency to motion; theſe are actual MOTION, and IMPULSE without motion: in each of theſe conditions bodies are active or endeavour to move, the endeavour in one caſe being manifeſt by the actual motion, in the other by the force neceſſary to reſtrain the motion; and a thing cannot be conceived to have a tendency to motion without being in one, or the other, of theſe conditions.

. P   Therefore

J   Therefore whatever tends to move, muſt either actually move, or it muſt impel, without actually being in motion.

K   But the ACTIVE SUBSTANCE cannot impel without being in motion.

L   IMPULSE is the action which happens between ſolids, in contact, at their ſurfaces, when one impedes the others motion, on account of their mutual reſiſtence to penetration.

M   But the ACTIVE SUBSTANCE is not ſolid, and does not reſiſt penetration (F 85); it is therefore, incapable of impelling or of ſuſtaining impulſe; it cannot, like bodies, reſt in certain portions, preſſing upon ſurfaces of other portions, either of its own ſubſtance, or of matter; becauſe, in every part it is permeable and yielding, and no ſurface reſiſts another.

N   Therefore, the ACTIVE SUBSTANCE tending to move, and being incapable of having its motion impeded by impulſe, does actually and continually move.

O   And ſince the ACTIVE SUBSTANCE continually moves, and is never diſpoſed to reſt, nor is capable of impulſe, it cannot be active in the

the manner of reſt, nor of impulſe, but muſt be active in the manner in which it exiſts, that is by its motion.

Since, alſo, it is eſſential to the ACTIVE SUB- STANCE to be active, and motion is the manner in which it is active, MOTION IS ESSENTIAL TO THE ACTIVE SUBSTANCE. *P*

The ACTIVE SUBSTANCE is *active* of itſelf, and moves of itſelf; but in order that it may *act*, ſome other thing, upon which it may produce a change, is neceſſary. *Q*

For whatever ſuffers an action, receives ſome change. *R*

The active ſubſtance in acting on ſome other thing, muſt impart and unite itſelf thereto, for its *action* is communicating its *activity* (c 104); but it cannot communicate its activity without imparting its ſubſtance, becauſe it is the ſubſtance alone which poſſeſſes activity, and the quality cannot be ſeparated from its ſubſtance. *S*

THEREFORE THE ACTIVE SUBSTANCE ACTS BY UNITING ITSELF WITH THE SUBSTANCE ON WHICH IT ACTS. *T*

The

U   The union of this fubftance with bodies is not to be conceived of as a junction of fmall parts, intimately blended together, and attached at their furfaces; but as an entire diffufion, and incorporation of one fubftance with another in perfect coalefcence.

CHAP. III.

On the different Origins and Circumftances of Activity in Bodies, apparent from Obfervation.

A   SOME of the apparent ORIGINS of activity in bodies are;

B   VOLITION; this appears from the actions of the mufcles, which are fubfervient to our wills, and the motion of the limbs confequent thereon:

C   IMPULSE; as appears in the motion, or action confequent on impulfe:

D   FIRE; as appears in the expanfion of air, and other bodies by its means:

ELECTRICITY;

ELECTRICITY; as appears in its attracting and E
repelling light bodies, and shattering and destroy-
ing larger ones, as in the effects of lightning.

The above origins are EXPERIMENTAL CAUSES, F
being events observed to precede, and be connected with motion as their effect.

Matter is active under certain CIRCUMSTAN- G
CES, when NO CAUSE OR ORIGIN is apparent: we
can, in these cases, only mark the CIRCUMSTANCES
which distinguish one activity, of one denomination, from another activity, of another denomination; but we are not to mistake, as signifying a cause, the term which serves only to
designate the species of activity.

Among the different activities, of which we H
know no EXPERIMENTAL CAUSE, or, can by
our senses, discover no ORIGIN, we may enume-
rate

Those of certain organized systems which we I
denominate LIFE; as in living animals and
vegetables.

Those whereby bodies merely approach to- K
wards or recede from others: as bodies approach to the earth; iron to the magnet; as
two

two fimilar poles of a magnet recede from each other. The actions on which thefe motions depend are called in general ATTRACTION and REPULSION; the particular attraction towards the earth is called GRAVITY; and of the fmall parts of bodies to each other, COHESION.

L   Thofe whereby bodies, whofe figures are changed, reftore themfelves to their former figures, termed ELASTICITY.

M   We might diftinguifh other varieties of activity, were it important to do fo; but all of them belong to one or the other of thefe two heads; thofe where an EXPERIMENTAL CAUSE is manifeft, and thofe where NO CAUSE AT ALL is difcovered by the fenfes.

N   And here it is proper to notice a prevailing error, in making words ftand for names of CAUSES of motion, which, in fact, can only ferve to diftinguifh the SPECIES, there being no CAUSE difcovered. The motion of the fmall parts of bodies approaching each other to cohere; the motion of the iron to the magnet; the motion of bodies towards the earth, may, very well, be diftinguifhed from each other, and from other motions. If we called them all, in general, attractive motions; one,

attraction

attraction of cohesion; an other, attraction of magnetism; the third, attraction of gravitation; to these names, as distinguishing species and differences, no objection could arise: but when we change the mode of expression, and say, attraction is the CAUSE of cohesion, of magnetism, and of gravity; the terms serve for a very different, and a very exceptionable purpose; they are no longer names of things perceived, to wit, motions or tendencies, but they become names of imaginary, and unknown things; of things unperceived by sense, and unconceived in the mind. When, for example, it is said the parts of bodies cohere *by* an attraction, what idea have we of the thing signified by attraction? If we say bodies tend to the earth *by* their gravity, what idea have we of this gravity? But if we define the term gravity to signify the actual tendency of bodies to the earth, and say they gravitate *by* attraction, then what do we mean, different from the fact of gravity, by attraction? We must give a name to every known fact, whether cause or effect; but we ought not to give a name to a cause while it is unknown either as experimental or rational; for this is to use a word without an idea, to perplex the mind, and retard the progress of knowledge, by a substitution of its semblance for the reality.

In

O   In all thofe motions, or tendencies, termed ATTRACTIVE, we fee no event or circumftance precede the motion or tendency; nothing to which, as a caufe, we can refer the effect; the EFFECT therefore is manifeft, but the CAUSE hidden.

P   In *no* example of activity in matter is the RATIONAL CAUSE manifeft. Among thofe activities where we can fee an EXPERIMENTAL CAUSE, fuch as volition, impulfe, fire, electricity, we do not perceive the neceffary connection between caufe and effect; we do not *underftand* the facts. In the other motions, we fee, as was faid, NO CAUSE AT ALL; the RATIONAL CAUSE of all activity in matter, remains therefore, to be inveftigated.

Q   Whenever a body is fubject to a caufe of motion, or in circumftances wherein, if left to itfelf, it would be moved, and is prevented from motion, it neverthelefs becomes active in the direction of the caufe, and exerts an impulfe on whatever impedes its own motion.

CHAP.

C H A P. IV.

Of Activity in the General, confidered as appertaining to Bodies.

WE propofe, now, to fpeak of bodies as A whole and entire maffes, in which fenfe they are, of themfelves, inactive (M 18); *i. e.* without any tendency to motion, and confequently, whatever activity they poffefs muft be derived from fome external fource (A, &c. 108).

Bodies (A 104) have two forms of activity, B MOTION and IMPULSE (105); a tendency to motion is common to both of thefe forms of activity; and all activity is, in this refpect, fimilar to itfelf, having only differences of direction and degree.

Therefore, one caufe only, differently circumftanced, is to be affigned for all activity in bodies (D 103).

The internal activities (F 24) among the parts of D bodies, as well that whereby the primary parts are hard and refifting, as that whereby they cohere together,

Q

together, may be called the CONSTITUENT ACTI-
VITIES; the former the CONSTITUENT ACTIVITY
of MATTER, becaufe a primary part is the firſt
and moſt ſimple form of matter; the latter the
CONSTITUENT ACTIVITY of SPECIES, becaufe
the manner in which parts are arranged by
theſe actions, may be conceived to conſtitute
the fpecific differences of bodies.

E From thefe CONSTITUENT ACTIVITIES a
CONSTITUENT ACTIVE SUBSTANCE is inferred,
as the fubſtratum of matter (1 59).

F The CONSTITUENT ACTIVITIES, by their
equilibrium, leave the WHOLE of every mafs
INACTIVE, quiefcent, paffive; not repugnant to
motion, nor refiſting, but a mere privation of
activity (A 113).

G Whatever other ACTIVITIES, befides the CON-
STITUENT, are found in bodies, thefe, not being
innate, nor arifing from an internal fource, nor
neceffary to the exiſtence of bodies, but originating from without, and fuperinduced on them
fubfequent to their formation, may be called
ACCESSORY ACTIVITIES; of this kind is the
activity of gravity, of magnetifm, of animal
volition, of impulfe, and all others not included
under the CONSTITUENT ACTIVITIES.

THE

THE ACCESSORY ACTIVITIES ARE SUPER- H
INDUCED UPON BODIES BY THE ACCESSION OF
A NEW ACTIVE SUBSTANCE, DISTINCT FROM
THE CONSTITUENT (E 114).

For bodies exist INACTIVE by the equilibrium of the CONSTITUENT ACTIVITIES (F 114); if, therefore, bodies are active, as in the accessory activities (G 114), since they are not active by their innate or CONSTITUENT ACTIVE SUBSTANCE, they become so by a new and superadded ACTIVE SUBSTANCE.

The new ACTIVE SUBSTANCE, which induces I
the ACCESSORY ACTIVITIES may be called the
ACCESSORY ACTIVE SUBSTANCE.

There are, therefore in a body, two forms of K
the ACTIVE SUBSTANCE, one is the essential ACTIVE SUBSTANCE which constitutes the body, in a manner to be explained(a); the other is the occasional ACTIVE SUBSTANCE by which a body is, in any way, made active, as in gravity or projection.

THEREFORE ALL THE ORIGINS OF ACTI- L
VITY ENUMERATED (A 108) SERVE TO IMPART
THE ACCESSORY ACTIVE SUBSTANCE TO BODIES,
AND WHERE ACTIVITY IS WITHOUT ANY MANI-

(a) See Chap. VI.

FEST ORIGIN (H 109) THE ACTIVE SUBSTANCE IS DERIVED FROM AN INVISIBLE SOURCE.

M THE ACTIVE SUBSTANCE BEING IMPARTED TO A BODY, PENETRATES THE MOST SOLID OR RESISTING PARTS, AND DOES NOT RESIDE IN THE PORES, WITHOUT, AND AT THE SURFACES OF THE SOLID PARTS. For the activity is imparted to the body itself, and not to its pores which are not parts of the body; therefore, if the ACTIVE SUBSTANCE remained within the pores, the cause would not, as is required by the second rule, be present with its effect; but the cause would be in one place, and the effect in another, which is impossible.

N Therefore, we are to confider of the communication of the ACTIVE SUBSTANCE, as of its forming a perfect union of its whole fubftance with the fubftance of the body; the two becoming, during this union, one fame fubftance and nature; and hence it is, that the body becomes active by the fuperaddition and union of this foreign active nature with its own inactive mafs.

O The active fubftance pervades and penetrates bodies without refiftence; for the refiftence of cohefion is only a refiftence of parts to feparation,

tion, and therefore to the interpofition of other *matter* which would feparate the parts; but it refifts not to an *immaterial* fubftance, which enters, and fills the mafs without feparating its parts, itfelf having no cohering or refifting parts.

BODIES ARE ACTIVE ALWAYS IN THE SAME P
DIRECTION IN WHICH THE ACTIVE SUBSTANCE
MOVES.

For activity as a caufe in one direction, cannot involve activity in another direction, as an effect (A 102).

THE QUANTITY OF ANY ACCESSORY ACTI- Q
VITY IN A BODY IS PROPOTIONAL TO THE
QUANTITY OF THE ACCESSORY ACTIVE SUB-
STANCE INDUCING THE ACTIVITY IN THE
BODY.

For the ACTIVE SUBSTANCE *fuperadded*, cannot *impart* more or lefs activity than is effential to it.

There are two ftates of inactivity, or inactivity R may depend on two different circumftances, the *privation* or the *equilibrium* of the ACTIVE SUBSTANCE. If a body has an ACCESSORY ACTIVITY, as, an impulfe from a man's hand, when that impulfe ceafes, the body will become, as before, inactive, and this inactivity may be confidered

as

as depending on the *privation* of the impulſe, or it may be conſidered as depending on the *equilibrium* (of the conſtituent actions (F 114), as if no impulſe had been given to it. Again, if a contrary and equal impulſe be given from another hand, the body will alſo become inactive, as if no impulſe were given to it, by the *equilibrium* of the two impulſes.

s - WHEN AN ACCESSORY ACTIVITY IN A BODY CEASES, EITHER THE BODY IS DEPRIVED OF ITS ACCESSORY ACTIVE SUBSTANCE, OR HAS RECEIVED A NEW AND EQUAL QUANTITY OF ACCESSORY ACTIVE SUBSTANCE, MOVING IN A CONTRARY DIRECTION.

- The body loſes its ACCESSORY ACTIVITY either by the *privation* or the *equilibrium* of the ACCESSORY ACTIVE SUBSTANCE (U 117); for ſince the quality of activity has ceaſed, either by the ſecond rule, the ſubſtance poſſeſſing that quality has quitted the body, or by the fifth, the quality has been deſtroyed by the union of its contrary.

T   In order the better to underſtand the ſeveral ways in which matter may be ſaid to be active or inactive, the following conſiderations may be uſeful.

Activity

Activity in bodies may be confidered as the
activity in parts, or in the whole of any finite
mafs. Cohefion is an action among the *parts* of
a body; gravity, or projectile impulfe, is an
action of the *whole* mafs of any body; but if
we confider gravity comprehenfively as in the
whole mafs of earth, and bodies, not as *wholes*,
but as *parts*, forming collectively the whole earth,
then gravity is to be confidered not as an activity
of the *whole* earth, as an unit, but of its *parts*
numerically. If we confider the earth as a whole
mafs, its motion in its orbit may be called the
activity of a *whole* body; but if we have regard to the folar fyftem as a whole, of which
the earth forms a part, then we are to confider
the planetary motions as activities of the *parts*
of that inactive *whole*. Thus every mafs may
be confidered in the two characters, of a *whole*,
and of a *part* of fome greater *whole*; as the
former *inert*, as the latter *active*.

The activity which bodies have, confidered
as *wholes*, are internal. That which they have
confidered as *parts* of a greater whole, are derived from without: thus a ftone, or any portion of the earth, conftituted a diftinct mafs
by the cohefion of its parts, has, as a whole,
no activity proper to it but this activity of parts
which

which conſtitutes it; gravity, or any other activity is not eſſential to its exiſtence as a ſtone, but is ſuperadded. The earth, regarded as a whole, has its activity of gravity within and eſſential to it; for it is formed a whole only by its gravity; but the activity whereby it moves in its orbit is without, and not eſſential to it as a globe of earth. The activity of coheſion is eſſential to matter, as matter; but the activity of gravity is not eſſential to matter as ſuch, but to the earth as an earth. The activity of the earth in its orbit is not eſſential to the earth as ſuch, but it is eſſential to it as a part of the ſolar ſyſtem.

y In reſpect to its own internal activities, EVERY WHOLE may be ſaid to be *inactive*, becauſe the reſult of the internal activities of the parts is an equilibrium: the whole is active inaſmuch as all the parts are active, but not active *as a whole*, becauſe from all the activities of parts no one activity of the whole reſults.

z In this ſenſe we may ſay that every PRIMARY ATOM, every MASS OF MATTER, THE EARTH, and THE SOLAR SYSTEM, is INACTIVE, becauſe each of theſe reſpectively is *in equilibrio* relative to the activities of its own parts.

But

But every whole is active as a whole, relatively A to fome greater whole, of which it forms a component part (u 119).

In this fenfe a PRIMARY ATOM, a MASS OF B MATTER, and the EARTH are ACTIVE. The atom is active relatively to the mafs it helps to conftitute, by its cohefion to other atoms. The mafs is active relatively to the earth, of which, as a greater whole, it, by its gravity, forms a part. The earth is active in relation to the folar fyftem, and it is poffible, the folar fyftem may be active, in regard to the larger fyftem of the univerfe.

The above diftinctions and explanations concerning action are important to be well underftood, and kept in view.

CHAP. V.

Of Activity in Bodies, as distinguished into Motion and the two Forms of Impulse, Pressure and Percussion.

A A STATE of activity is superinduced upon bodies, of themselves inactive, by the union of the ACTIVE SUBSTANCE, with the substance of the bodies (N 116).

B The state of activity of bodies appears under two forms, motion and impulse: so the activity of gravity is manifest in a falling body by its motion, in a body at rest by its impulse.

C There are two forms of impulse, *pressure* and *percussion*; a body at rest presses by its gravity, but when a falling body, or a projectile undergoes the change from motion to rest, it percusses, strikes, or impinges; which terms may be used to signify the same thing.

D Thus activity in bodies has three different forms, motion, pressure, and percussion; but the

the ACTIVE SUBSTANCE has only one manner of being active, that is, by motion (o 106); and one mode of acting, by uniting itfelf to that on which it acts (T 107).

It remains to be inveftigated how the ACTIVE E SUBSTANCE can by its *motion* alone, induce on bodies when united to them, the three varieties of activity, *motion, preffure,* and *percuffion,* and on what relative circumftances of this SUBSTANCE to bodies thefe diverfities depend.

When any quantity of acceffory ACTIVE SUB- F STANCE (I 115) is imparted to a body, the body acquires a ftate of acceffory activity, or an activity not proper to it, but induced upon it.

This activity will manifeft itfelf either by the G motion of the body, by its preffure, or by its percuffion on fome other body.

If the activity of the body (G 123), be mani- H fefted only by its motion, the body may be faid to be in a ftate of activity only, but not to act. It is in a ftate capable of action, but has no fubject on which to exert an action (c 104). This ftate of a body can only be conceived of as exifting in a motion in an unrefifting fpace.

R 2

A body

ɪ   A body may be in a state partly of motion, and partly of impulse; as when moving in any resisting medium. In this state the body is partly active, without acting, and partly acting. The *activity* of the body is as much as it either moves or impels the medium, and is equal to the sum of the motion and impulse; but the *action* of the body is equal to the impulse only, since the action can be as much only as there is a subject to sustain the action. A body may have the same *activity* in vacuo, in a rare medium, or in a dense one; but the *action* will be different in each of these cases; in the vacuum there can be no action; in the rare medium there will be less action than in the dense; in the dense more; and most of all, if instead of a penetrable medium there were an unyielding obstacle, to receive the *action* of the whole *activity* of the body.

ĸ   When in the former part of this work it was said, that motion implied action (ᴘ 48), we were not prepared to make the distinction, here necessary to be attended to, between activity and action. We spoke of action there in its general sense as opposed to inertia; motion could not be admitted to depend on inertia, but was proved to be preserved by an active cause, and we did not scruple to call it an action, till now that our further

further progress leads us to more accurate discriminations. This will be the case with other terms which relate to the main business of our inquiries. We must begin with using them in the accustomed sense; but as the subject opens and becomes more clear, new distinctions and new definitions will be necessary. The doctrine that a body persevered in an uniform motion in a right line by its inactivity, was combated and shown to be an error, and the whole motion, beginning, middle, and end, proved to be one continued action; but now, that we distinguish between action and activity, it will be more just to say, that the whole continued motion is one continued activity, and that action exists only when the body is acquiring motion by the communication of the ACTIVE SUBSTANCE (L 115), or when the motion is diminishing or ceasing by the impulse of the body on obstacles, which obstacles themselves acquire activity by that impulse.

From what has been said we may learn that action is proportionable always to the activity, and, if I may so say, the passivity, compounded together; for the activity cannot act without something on which to act, that is to unite itself to (N 116), nor can there be any action from the patient alone without something active. The

same

same activity may, we have seen, have any quantity of action, from nothing to its whole quantity (I 124), or till the whole ACTIVE SUBSTANCE is united with the patient which receives its action.

M   BODIES BY IMPULSE ON OTHERS LOSE THEIR ACTIVITY IN PROPORTION TO THE IMPULSE.

This is matter of observation. Bodies which move in rare *mediums* move for a longer time, because they impel less; and if they meet any obstacle capable of receiving the whole activity at once, in the impulse of the body, the body suddenly loses its whole activity and comes to rest.

N   Hence, a body by acting loses the activity which enabled it to act; for only the impulse of the body is its action; its motion is not its action but its activity; by impulse it loses its motion, that is, by acting it loses its activity.

O   BODIES WHICH SUFFER IMPULSE ACQUIRE ACTIVITY IN PROPORTION TO THE IMPULSE.

This also is matter of observation.

P   The ACCESSORY ACTIVITIES of bodies (G 114) as explained to be those which are not proper to bodies, but derived, have been showed to be be induced upon them by the acceffion and union of the ACTIVE SUBSTANCE with them (H 115).

When any ACCESSORY ACTIVITY in a body Q ceafes, it has either received a new and contrary activity, or is deprived of its acceffory ACTIVE SUBSTANCE (S 118).

IN IMPULSE THE ACTIVE SUBSTANCE PASSES R OUT OF THE IMPELLING BODY INTO THE BODY IMPELLED.

Bodies in motion are active (P 48), and activity confifts in the prefence of the ACTIVE SUBSTANCE (I 59), and by impulfe bodies lofe their activity (M 126), therefore they lofe their ACTIVE SUBSTANCE, and the lofs is proportional to the impulfe (M 126). Bodies impelled acquire activity (O 126), therefore acquire ACTIVE SUBSTANCE, and they acquire proportional to the impulfe (O 126); now the ACTIVE SUBSTANCE loft by the impelling body ought to be concluded to be that found in the other, becaufe there is no other receptacle than the impelled body to which the fubftance parted from can be traced, nor any other fource than the active body from whence that which is found can be derived; therefore in impulfe the ACTIVE SUBSTANCE ought to be concluded to pafs from the impelling body into the body impelled.

s THE CONTINUED MOTION OF A BODY DEPENDS UPON THE CONTINUANCE OF THE ACTIVE SUBSTANCE WITHIN THE BODY.

For the motion of the body depends on its activity, and its activity depends on its union with the ACTIVE SUBSTANCE; fo long therefore as the body continues moving, it continues active, and retains its ACTIVE SUBSTANCE.

T THE MOTION OF A BODY IS PRODUCED BY THE MOTION OF THE ACTIVE SUBSTANCE IN UNION WITH THE BODY.

It being evident that fince the ACTIVE SUBSTANCE itfelf does always move, whatever it is united to will be moved along with it, if no obftacle prevent.

Otherwife, fince the ACTIVE SUBSTANCE is itfelf active only by its motion (o 107), it is only by its motion that it can render bodies active, or produce motion in them when united with them.

U ACTIVITY IS, IN IMPULSE, COMMUNICATED FROM THE AGENT TO THE PATIENT BY THE FLOWING OF THE ACTIVE SUBSTANCE OUT OF THE AGENT INTO THE PATIENT.

It

It has been proved that in impulse the ACTIVE SUBSTANCE does actually pass from the impelling to the impelled body (R 127); consequently activity is communicated by this means.

For the communication of an effect involves, by the second rule, the communication of the cause, otherwise the cause could not be present with the effect.

The flowing of the ACTIVE SUBSTANCE is a sufficient cause of the communication of activity, and no other rational cause can be assigned; this, therefore, ought to be considered as true, as well as sufficient.

In mere motion, the body moved is the patient, x and the ACTIVE SUBSTANCE the agent. In impulse, the body in motion may be considered as an agent, as it is made active by its ACTIVE SUBSTANCE.

Impulse may happen when the active body or y agent meets with another body, either at rest, or moving in some other direction than the direction of its own action, or moving in the same direction with its own, but with less velocity. But for greater simplicity, we will consider the impelled body, or patient, as always at rest, and in this light, as a patient, it ought to be considered; for if it be moving in a direction con-

S trary

trary to the agent, itself becomes also an agent. If moving in the same direction with less velocity, the impulse is the same as if the patient were at rest, and the agent were only moving with its excess or difference of velocity.

z   Therefore the patient as such, is to be considered as always at rest, and as inactive; consequently, as having no tendency to motion.

A   AN INACTIVE BODY BEING IN A PART TO WHICH AN ACTIVE BODY TENDS TO MOVE WILL BE AN OBSTACLE TO THE MOTION OF THE ACTIVE BODY.

For in order that the active body should move, either the inactive one must also move, so as not to impede it, or the active body must pass through the inactive one; but the inactive body cannot move, because it is inactive; and the active body cannot pass through it, because of the cohesion of the parts of the inactive mass; therefore, until one of those impediments are removed, it is impossible the active body should move.

B   THE ACTIVE SUBSTANCE UNITED WITH AN ACTIVE BODY OPPOSED BY AN OBSTACLE WILL FLOW OUT OF THE ACTIVE BODY INTO THE OBSTACLE.

The

The obſtacle, by its coheſion, reſiſts to the active body only (c 130), but does not reſiſt the admiſſion of the ACTIVE SUBSTANCE (M 116); the obſtacle, therefore, impedes the body only, but not the ACTIVE SUBSTANCE; and the ACTIVE SUBSTANCE continuing to move, leaves behind it the body to which it belonged, and, finding the obſtacle in its way, readily enters it.

It has been already proved by inference from the effects, that the ACTIVE SUBSTANCE does paſs out of an impelling body into one impelled. Here it is proved from the nature of the ACTIVE SUBSTANCE, as motive and unſolid, and the nature of matter as quieſcent and refiſting penetration by matter, that this muſt neceſſarily happen. C

THE OBSTACLE BECOMES ACTIVE BY THIS ACCESSION TO IT OF THE ACTIVE SUBSTANCE (II 115). D

WHEN A SUFFICIENT QUANTITY OF ACTIVITY IS THUS INDUCED IN THE OBSTACLE, IT WILL BE CAPABLE OF MOVING, SO AS NOT TO IMPEDE THE OTHER ACTIVE BODY, AND WILL BE NO LONGER AN OBSTACLE, UNLESS ITSELF BE ALSO IMPEDED BY ANOTHER OBSTACLE. E

WHILE

F   WHILE THE ACTIVE SUBSTANCE IS FLOWING OUT OF THE ACTIVE BODY INTO THE OBSTACLE (D 130), THE ACTIVE BODY WILL PRESS ON OR IMPEL THE OBSTACLE.

While the ACTIVE SUBSTANCE is yet within the body, although flowing through it, it does not ceafe to impart to the body its own nature, nor can the body ceafe to be active becaufe not yet deprived of the ACTIVE SUBSTANCE; therefore during its paffing out of the body, fuch portion of the ACTIVE SUBSTANCE as is yet within, is urging and difpofing the body to move, in like manner as if the ACTIVE SUBSTANCE were CONTINUING in the body (S 128), and the body being thus urged to move, but impeded from moving, preffes or impels the obftacle.

G   We fee here an obvious explanation of impulfe; it confifts in the flowing of the motive fubftance from a fource into a receptacle. In order to this, an *obftacle* muft impede the *active body*; for otherwife the ACTIVE SUBSTANCE will carry the body along with it, and not quit it. The *obftacle* muft receive this SUBSTANCE as it efcapes from the *active body*, becaufe it is in its path; and the ACTIVE SUBSTANCE, while quitting its firft body, ftill urges it to move, and caufes it to impel.

Although

Although the exiftence of the ACTIVE SUB- H STANCE had not been eftablifhed on any previous grounds, the phenomenon of communication of motion by impulfe does alone afford a philofophical proof of its reality.

Becaufe if a fact admits of only one explanation, and that be intelligible, analogous to the fact, agreeable to the common order of things, and complete, that explanation ought to be confidered as true, although it may be founded in hypothefis.

For it being the end of philofophy to explain K facts, it would not be confiftent with that end to reject an explanation, the only one which can be given, and one intelligible, analogous to the fact, agreeable to the common order of things, and complete, becaufe it refted on hypothefis; fince, rejecting this, we muft remain in ignorance, and we ought to prefer imperfect information to abfolute ignorance. But further, fince no other explanation can be given, we ought to confider the hypothefis as proved, and no longer hypothetical; fince where there is but one natural means of producing an effect, nature muft operate by that means; where we find but a fingle road we are certain we cannot err in our choice.

Now

L   Now if we reject the doctrine of communication of some motive virtue, between bodies in impulse, we can no otherwise explain the parts of the phenomenon, nor the final effect : for example, how the agent has its motion diminished; how the patient has motion given to it; why the quantities of these motions are equal; why the changes happen at the same time; and why they depend on the connection of the two bodies.

M   But when motion is lost out of one body, and a similar motion is found in another body so connected with the former as to allow of communication, and the quantities lost and found are equal, and the appearance in one body answers in time to the disappearance in the other, all of which may be said of the motion in impulse; the supposition of communication of a motive virtue is intelligible, analogous to the appearances, agreeable to common observation, affords a complete explanation, and ought therefore to be considered as true: but this explanation being admitted, the existence of the ACTIVE SUB-STANCE must also be admitted on this ground alone; for that must necessarily be A SUBSTANCE which can pass, or be communicated from one body to another.

N   But if any should refuse to admit a new

principle

principle of such importance unless it can be supported by facts; it is replied, that, in this case, facts cannot be adduced in direct proof; ought not to be expected; and that the rational evidence before us, is already stronger than any which could arise from sensible appearances alone, were such to be produced, without the same rational demonstration. So great in the present day is the predilection for facts, that men seem to consider truth in no other light than as tangible, visible, or manifest to some sense. If we could in impulse, discern something passing from one body to the other, this appearance, without the rational evidence adduced, would not be a sufficient ground on which to establish a principle, like that of the ACTIVE SUBSTANCE. We might suspect, in this case, an illusion of vision; and even, if we could not doubt of the fact, still, we should want assurance that this visible fluent was itself the active principle of impulse; or rather we should be assured that it could not be; for were it visible it would be material, if material, inactive (v 120), and incapable of being a principle and source of activity. Facts are the sensible basis from which we infer our principles, the truth of which appears in the satisfactory explanation of facts which they afford; but so far is it from being possible, that any experiment should discover this agent,

the

the ACTIVE SUBSTANCE to the fenfes, that the being difcoverable by the fenfes would be itfelf a proof that the thing fo difcovered was not this agent in natural effects.

o   Hitherto we have fpoken of impulfe in the general only, we are yet to confider it as it is diftinguifhable into its two forms of preffure and percuffion (E 123).

P   Preffure is the impulfe that an active body exerts upon another, with which it is in contact. Pulling may alfo come under the head of preffure, as a modification of it.

Q   Percuffion is the impulfe that an active body exerts on another at the time of coming in contact with it.

R   Or, they may be thus diftinguifhed: Preffure is the impulfe of a body having more activity on another having lefs, both having equal velocities in the fame direction, or both being at reft. Percuffion is the impulfe of a body having more velocity on another having lefs in the fame direction. A man *preffes* a body that he pufhes before him, both moving with equal velocities; but the man having more activity, or tending to move fafter than the body. A ftone projected

percuffes

percuffes or ſtrikes againſt any fixed body, the ſtone having the greater velocity, before colliſion.

It has been ſhown that impulſe in the general, s whether of preſſure or percuſſion, produces a flowing of the ACTIVE SUBSTANCE out of the agent into the patient, ſo long as the impulſe continues (R 127).

This circumſtance being common to both T modes of impulſe, it remains to inveſtigate in what peculiarities of the flow conſiſt the difference between the two modes of preſſure and percuſſion.

Theſe we muſt ſeek by attending to the U differences in the faƈts themſelves.

Preſſure is an impulſe, which may be con- X tinued indefinitely, of an aƈtive body on another with which it is in contaƈt (P 136); again, in preſſure, the velocities of agent and patient are equal, that is, both may be conſidered as at reſt (136).

Percuſſion is an impulſe, which cannot be Y continued, of a body at the time of coming in contaƈt with another: alſo, in percuſſion, the

T agent

agent has more velocity, before collision, than the patient; therefore, percussion is to be considered as the act of a body in motion, and prepared for its action, by its motion (R 136).

z   Since pressure is an impulse that may be continued, the flowing of the ACTIVE SUBSTANCE is continued, during the pressure.

A   Since in percussion the impulse is transitory, the flowing of the ACTIVE SUBSTANCE is also of short duration.

B   It follows from hence, that if the same quantity of motion be imparted, by a pressure of some continuance, and by a percussion, since, in either case, the same quantity of ACTIVE SUBSTANCE will be transmitted or flow from the agent to the patient, and in the case of percussion it will flow in a shorter time; the flow must be more rapid and violent in percussion than in pressure.

C   Since, during any pressure, the ACTIVE SUBSTANCE continually flows out of the pressing body, either that body itself must be the source of this flow, or the body must be the channel through which the ACTIVE SUBSTANCE flows from some other source.

But,

But, it appears from obfervation, that a body D does not prefs, unlefs it be fubjected to fome of the circumftances or origins of ACCESSORY ACTIVITY (I 115); either the regular action of gravity, or fome accidental action, as a man's hand, a bended fpring, &c.

Hence it appears, that a body preffes by means E of an ACCESSORY ACTIVE SUBSTANCE (I 115), entering from without and paffing through the body, and caufing it to prefs as it is flowing through it, in the manner explained (F 132).

And this alfo appears, becaufe the preffure of F a body is in fome cafes, as in gravity perpetual.

For if the ACTIVE SUBSTANCE originated in G the preffing body, its quantity muft have fome limits, and, after a certain time, greater or lefs, the ACTIVE SUBSTANCE in any body continuing to flow out would be exhaufted; and before it was altogether exhaufted, it would be continually diminifhed, and the preffure would continually decreafe till it finally ceafed; but thefe do not agree with the facts.

In other cafes the time of preffure depends H on external circumftances, and not at all on any thing in the nature or magnitude of the body;

as

as when a man, or a fpring, caufes a body to prefs; therefore, the bodies, in no cafe, prefs by an inherent activity, but by a fuperadded one.

I   A percuffing body, on the contrary, is itfelf the fource of the ACTIVE SUBSTANCE, which flows out of it on collifion.

K   For a percuffing body is in motion previous to percuffion, and the previous ftate of motion is neceffary to fit the body for percuffion (Y 137).

L   Now, a body during motion, is not neceffarily fubjected to any of the origins or circumftances of acceffory activity (A, &c. 108); a projectile, for example, has at the time it received the impulfe which made it a projectile, received the ACTIVE SUBSTANCE from fome origin, as elafticity, or animal volition; but the body, flying from this fource, has no longer any connection with it, nor is fubject to its influence. The ftone impelled by a man's arm, has no longer any fubjection to the arm when it is paffing through the air; it therefore continues to move by a retained and inherent activity.

M   Again, in *preffure*, the fmaller of two bodies, both at reft, may prefs more, becaufe the preffure depend not on the bodies; but if two bodies are

are *moving* with equal velocities, the greateſt maſs will always have a proportionally greater capacity of percuſſion; therefore, the percuſſion is dependent on the maſs, and is performed by an inherent activity, imparted to the body, when it was put in motion.

It has alſo been ſhown that the motion of a N body depends upon the continuance of the ACTIVE SUBSTANCE within the body (s 128).

We may diſtinguiſh the ſtates of the ACTIVE O SUBSTANCE, as it is in a moving body, and in an impelling body, into the STATE OF RETENTION and the FLOWING STATE. During motion, the ACTIVE SUBSTANCE continues in the body, and this may be called its RETAINED STATE; but it is to be obſerved, that by retained, it is not meant that the ACTIVE SUBSTANCE is held or confined by the body, but only that it remains within it. In impulſe, the ACTIVE SUBSTANCE being paſſing or flowing out of the impelling body, this may be called the FLOWING STATE of the ACTIVE SUBSTANCE. Here again it is to be obſerved, that the term flowing is employed only to ſignify the ſtate of the ACTIVE SUBSTANCE relative to the body; for in both ſtates the ACTIVE SUBSTANCE itſelf is moving, and may be ſaid always to flow; but in the ſtate

of

of retention, the containing body moves with it; in the flowing ftate the ACTIVE SUBSTANCE paffes out of the body, or flows through it.

P   A body in motion has received a certain quantity of ACCESSORY ACTIVE SUBSTANCE from the agent which gave it motion, and when it has gotten out of the reach of receiving any increafe, this quantity already poffeffed continues in the body, by its own motion, moving the body (T 128), but remaining at reft in and relative to the body, as the rowers of a veffel which give motion to the veffel, move together with it, but are relatively to the parts of the veffel at reft.

Q   The moving body is thus poffeffed, in itfelf, of a limited quantity of ACCESSORY ACTIVE SUBSTANCE, which, as it has received, it can part with again; and as it moves by the prefence of this fubftance, fo when it no longer poffeffes it, it ceafes to move.

R   Therefore, when an obftacle is oppofed to a body in motion, if it be of fuch magnitude as to receive no fenfible velocity from the blow, the obftacle will rob the moving body of its ACTIVE SUBSTANCE; for the obftacle obftructs the paffage of the body, but the ACTIVE SUBSTANCE

freely

freely pervading all bodies paffes on, out of the body, now at reft, which it before moved, and enters the obftacle.

Hence percuffion is tranfitory, becaufe there is no continued fupply of the ACTIVE SUBSTANCE, but as foon as all that was in the projectile has flowed out of it, there is no longer any flow, nor does the impulfe continue any longer.

A preffing body at reft continually fuffers the ACTIVE SUBSTANCE to flow through, and thus can never accumulate a quantity, and never increafes its own activity, fince it parts with its ACTIVE SUBSTANCE as faft as it receives it.

But a body, when it acquires motion, has accumulated a certain quantity of ACTIVE SUBSTANCE, which it retains. For in order to its acquiring motion, there is required the abfence of fuch obftacles as might impede the body; that is, as might ferve to receive and rob it of its ACTIVE SUBSTANCE. Two circumftances therefore, connected with each other, go to the motion. The abfence of an obftacle, and the confequent retention of the ACTIVE SUBSTANCE in the body; from thefe the motion of the body neceffarily follows (T 128).

Therefore,

x Therefore, the impulfe of percuffion is the impulfe of a body, which, having previoufly been in motion, has obtained an accumulation of ACTIVE SUBSTANCE, continuing or retained within the body.

Y At the time of the collifion, this accumulated quantity of ACTIVE SUBSTANCE then begins to become a flowing fubftance in the impinging body, becaufe then, and not before, it begins to efcape from that body into the obftacle.

z Thefe then are the feveral circumftances of the flow of the ACTIVE SUBSTANCE, on which the different facts which characterize preffure and percuffion depend, and which ferve to explain thofe facts.

A Percuffion is, when the ACTIVE SUBSTANCE, from a ftate of retention, changes to the flowing ftate, which is, when a body in motion firft meets an obftacle, and the motion is changing to an impulfe (Y 144).

B Preffure, is when the change from the ftate of retention to the flowing ftate, or from motion to impulfe, is completed, and the flowing ftate actually exifts.

The

The force or efficacy of percuffion or preffure C to produce any effect, confifts not in any ftate or difpofition of the body percuffing or preffing, but in the paffing of the ACTIVE SUBSTANCE out of this body into the patient, and imparting its activity to the parts, or the whole of thefe bodies, wherever it pervades.

Since percuffion confifts only in the change of D the condition of the ACTIVE SUBSTANCE from its retained to its flowing ftate, preffure muft fucceed to and terminate every percuffion; for when the ACTIVE SUBSTANCE, in an impinging projectile, has commenced its flowing ftate, it then goes on actually to flow out of the projectile; and during this actual flow, the body is preffing, the percuffion being over: but projectiles prefs for a very fhort and almoft infenfible time after they ftrike, becaufe they are foon deprived of their whole quantity of acceffory ACTIVE SUBSTANCE (R 142).

Bodies falling by their gravity, afford a good E illuftration of both thefe ftates. While a body is falling it continues expofed to the influence of gravity, that is, it is acquiring ACTIVE SUBSTANCE, but, meeting with no obftacle, the air not being here confidered, it parts with none, or retains all the ACTIVE SUBSTANCE it acquires,

U which

which therefore accumulates continually, and increases the celerity of the body till it meets with an obstacle or support, when the body, by a percussion, suddenly parts with this accumulated quantity of ACTIVE SUBSTANCE; after which it continues to press downwards, not with the same vehemence with which it is impinged by means of its accumulated activity, but with the equable activity of the ACTIVE SUBSTANCE of gravity continually passing, through the body, to the earth.

This subject of impulse will be further prosecuted in another place, and the various facts of pressure and percussion, be more largely explained. Thus far, the general principle elucidates the general facts: all impulse is, to appearance, the action of body at rest: thus, in pressure, the body is all along at rest; in percussion, the action begins when the motion begins to be impeded; but action can never be conceived as originating in quiescence, or as allied to rest; and hence the obscurity of all theory, in regard to impulse, when body alone is considered therein: but a knowledge of the unseen agent that pervades bodies and continually moves through their substance while the bodies are at rest, reconciles facts to reason, and reflects a light upon the inquiry, which could never have been derived from experiment.

CHAP.

CHAP. VI.

Of the Production of Matter from its immaterial Elements.

THE origin of that activity in bodies, termed accessory (G 114), although not innate, is no longer obscure, since there appears to be a substance in nature essentially active, uniting with filling, and freely pervading bodies, and thus inducing the accessory activities.

The varieties also of activities in bodies, as distinguished into motion and impulse (B 122), become obvious and familiar, as referable to the presence of the same cause, the ACTIVE SUBSTANCE, under different given circumstances. In the conceiving of these things there arises no difficulty; the existence of the ACTIVE SUBSTANCE itself being once comprehended, its effects in union with body, to render it active, and make it either move or impel, follow, by a necessary consequence, from the conditions supposed.

C   But it may feem more difficult to comprehend that matter confifts of immaterial elements; and that an unfolid and active being fhould, by any modifications whatever, become a folid and inactive mafs. This is a transformation that may, to many, appear not only problematical and difficult to conceive, but wholly impoffible, and implying contradictions, abfolutely and for ever irreconcilable.

D   But thefe difficulties, which appear to impede a tranflation of the fame fubftance from immaterial to material, and reciprocally from material to immaterial, arife only from erroneous preconceived conceptions of the nature of matter, and the way in which it differs from immaterial exiftence; and when thefe errors are corrected, the difficulty is removed.

E   If the nature of matter be a pure and effential *inertia*, an abfolute privation of active qualities; if it be, in every view, both as a whole, and in ts parts, in itfelf, and in its elements inactive; without doubt it will follow that a fubftance immaterial and effentially active can never by any modification become matter; fuch fubftance, and matter, can only be conceived of, as beings wholly unrelated; without any common property, without any intelligible means of union,

of

of attachment, of mutual influence, separated to
our conceptions by an immeasurable distance; and
their actual or supposed union in man's nature,
as a perpetual miracle.

But the nature of matter conceived, as it is,
at once active and inert, gives it a relation to
immaterial natures, by which the connection,
transformation, and union of the two, may also
readily be conceived.

When we no longer consider the solidity of
matter, as an inactive repletion of space with an
unknown substratum, substance, or essence;
with parts *held* together, in a manner incomprehensible, without *action*; hard, impenetrable,
and resisting power, but itself without power;
when in the room of these incongruous ideas,
and this contradictory language we substitute an
idea of matter as being composed of active
elements, by their own activity forming attachments, existing in union, and resisting separation;
and by the reciprocal and equal contrariety of
these activities, every union, thus formed, remaining in an equilibrium, and becoming inactive. This view of matter affords us an easy
solution of a problem, which, upon the other,
would have been for ever, as hitherto, inexplicable.

Considering

H   Confidering it therefore, as a truth already demonftrated, that the ACTIVE SUBSTANCE is the elementary principle, effence, or fubftratum of matter; fince thefe elements, and thefe alone refult from a decompofition, actual or ideal of matter; it is here to be inquired the manner how thefe elements form their compound, and by what modification the ACTIVE SUBSTANCE, retaining its own nature and effential properties, continuing immaterial, unfolid, and active, puts on at the fame time the form of matter, and becomes material, folid, and inert.

I   I am well aware that not only the common prejudices of mankind, but even thofe of philofophers, of almoft every defcription, will militate ftrongly againft the very principle of the inveftigation here propofed.

K   The production of matter, out of another fpecies of being, not material, is, as far as I know, a problem in philofophy altogether new, and which, fo far from having been already folved, has never yet been propofed.

L   In all men's refearches, analyfes, and divifions on the fubject of bodies, matter has been confidered as the firft genus. Neither chemiftry by its experimental, nor metaphyfics by its rational

tional investigations, have ever proposed any higher object than to discover those most simple parts of matter, which are the basis of its compound varieties. Nor was it ever conceived that these simple parts, or that the general idea of matter itself, abstracted from its various modifications and species, was yet a compound derivative being, resolvable into a more simple essence, and to be traced to a more remote origin in the series of natural causes and effects.

But if any one examines the grounds on which he concludes matter to be an original being, underived from any prior existence (except the FIRST EFFICIENT CAUSE); and wherein is the impossibility of conceiving matter to be formed, by natural means, out of an immaterial substance, he will probably find these notions to have no other foundation than prejudice; a prejudice indeed so early implanted, and so deeply rooted, as to influence the mind like the clearest demonstration or most evident truth.

Almost all philosophers have agreed that there is between material and immaterial beings, so wide a chasm, so perfect a discontinuity, so total a deprivation of connection, similitude or common properties, that no chain of reasoning could ever lead the mind from one to the other,

or

or enable it to conceive of any relation or mutual intercourse between them. Hence, although moſt ſpecies of bodies underwent an analyſis, matter itſelf was never conceived of as a ſubject of decompoſition, or as a compound confiſting of principles more ſimple than itſelf.

o When any compound is reſolved into its component principles, thoſe principles have ſeparately each a different nature from the compound which they form when united, and require each to be marked by a different name.

p We muſt be careful here to diſcriminate between the diviſion of wholes into parts, and the ſeparation of compounds into principles; a half, fourth, tenth, or any fractional part, has the ſame nature with the whole and differs only in quantity; for this ſort of diviſion has no reſpect to the qualities of things, but to their relative quantities only. But when a certain nature or quality belonging to any body, is produced from the union of ſeveral other bodies, each of different qualities, here ariſe a ground of diviſion of a different ſort, which has no regard to quantities, wholes, or parts, but merely to the qualities of things; and when a ſingle ſubſtance reſults from an union of two or more, the ſeparation of thoſe things, thus united, reproduces

the

the original and feparate fubftances, while it deftroys the compound, which exifted only in their union. Wine, *e. gr.* is a compound liquor, produced from an union of feveral other things, neither of which, of themfelves, are wine. On diftillation, a pure fpirit, water, and earth, are found to have been united in the compound wine. The feparation of thefe deftroys the wine, and gives rife to the diftinct exiftence of the fpirit, earth, and water. The feparation of thefe is altogether different from a divifion of the wine, as it refpects quantity; for we may divide a portion of wine indefinitely into fmaller quantities, while each divided portion fhall ftill be wine, becaufe it retains in union the three principles, water, fpirit, and earth.

We may, therefore, to preferve thefe different forts of divifions diftinct, call the divifion that has refpect merely to quantity, a feparation of parts; the other, which has refpect to qualities, a feparation of principles or elements; thofe more fimple things, which by union form a more compound, being ufually denominated the principles or elements of the compound.

The latter of thefe methods of divifion anfwers alfo to the terms analyfis, refolution, or decompofition. And this, chemiftry practically applies to to particular genera or species of bodies, as minerals, vegetables, spirit, oil, salts, &c. to discover the more simple principles of these compounds, and again to those principles, as themselves compounds, to arrive at their principles. But to matter in the general, as a solid extended substance, not distinguished into this or that particular kind, this sort of anylysis has never, that I know of, been applied. Matter has been conceived as being capable of a division only into *parts* continually less and less, each part still being matter, but not of a decomposition into *principles* of a different nature from matter.

5   If we consider matter in its general nature, as a compound formed of more simple elements (H 150), these elements must differ from matter as the elements of wine, spirit, earth, and water, differ from wine. These have not the qualities of wine, and therefore do not answer to the definition of the term wine; so in like manner the separate elements of matter, if there be any such, cannot answer to the definition of matter; they cannot separately have the properties of matter, because these properties arise from their union only: these elements therefore cannot be matter, and, not being matter, must necessarily be immaterial.

Since

Since therefore an analyfis of matter muft T needs have led to this conclufion, that matter was compofed of elements or principles, which were immaterial, and this conclufion again would neceffarily have eftablifhed a relation between material and immaterial being, directly contrary to the powerful prejudices entertained, that this relation was utterly impoffible, this preconceived error ftood in the way, as an effectual barrier againft fuch an inveftigation, and precluded in the very outfet, every advance in this path, towards any further difcovery of truth.

It would not perhaps be difficult to trace the U origin of this prejudice, in certain religious hypothefes. Pious men, being defirous to exalt the fpiritual nature of the human foul, and philofophy having taught them that matter was inert, dead, lumpifh, perifhable; they thought they could no way dignify the mind fo much as by denying to it every property belonging to body, and difclaiming all relation to it. When, however, philofophy fhall have more juftly developed the nature of corporeal being, and afferted its higher dignity, it may no longer be thought unworthy or incapable of a relation to mind, and immaterial beings, and of an union with them.

x   It is, however, not a little remarkable that this opinion of the diſtinct and irreconcilable natures, of body and mind could ſo long have ſubſiſted, and ſo generally have obtained, in contradiction to the moſt obvious fact that ever preſented itſelf as a guide to truth, the actual and intimate union of our own minds and bodies; and that men ſhould, contrary to all juſt methods of inveſtigation, deem a perpetual occurrence in the uniform order of nature, a conſtant miracle and violation of nature's laws, becauſe not reconcilable to their hypotheſes, when they ought rather, laying aſide hypotheſes, to have conſidered this very fact as part of the general ſyſtem from which the laws of nature were to be learned.

y   This analyſis of matter, has been done in the firſt part of this work, and the reſult has been, as it muſt be in every reſolution of a compound into its elements, that the elements are of a different nature from the compound; and conſequently that the elements of matter are not of the nature of matter, but immaterial.

z   By immaterial, I mean here, and wherever I uſe the term, only to exclude from a being the properties which conſtitute matter, but not to aſſert any poſitive qualities of that being;

there

there may be as great diverfities, for aught we know, among immaterial fubftances, as among material ones, although we have not names or ideas of many of thofe diverfities.

In confidering matter as porous, we concluded that there muft be an ultimatum of porofity, or certain parts void of pore, which muft be confidered as primary parts or atoms, whofe imperfect continuity forms the primary and fmalleft pores. A

It is generally agreed that all fenfible maffes of matter are porous, that is, are compofed of a number of the imporous parts (A 157), each individually, too fmall to excite any fenfation, adhering at certain points only, and at other points difcontinuous. B

There are, therefore, two diftinct facts in every fenfible mafs of matter, the production of which we fhall examine feparately, in the order of their occurrence; *firft*, the formation of a primary and imporous atom, or the moft fimple portion of matter; *fecondly*, the union of thefe atoms, fo as to form larger, and at length fenfible maffes: this latter effect is commonly called cohefion, and this word by fome may be thought to explain fomething of the fact; but to fay that c

the

the parts of bodies ſtick together by coheſion is only to inform us that the parts of bodies ſtick together by ſticking together.

D   It is proper here, previouſly to entering upon the inquiry how the primary atoms are formed, to recall the reader's attention to what has been ſaid reſpecting their nature, and in what reſpects the atoms which are here conſidered as the moſt ſimple parts of bodies, differ from thoſe which are exhibited to our imagination in the modern philoſophy.

E   The atoms of the modern philoſophy are conceived to have an extenſion, never preciſely aſſigned, but in ſome finite relation leſs than the leaſt viſible magnitude.

F   Theſe atoms are ſuppoſed to have the greateſt poſſible ſolidity; and this ſolidity is thought to conſiſt in a fulneſs of the extenſion of the atom, in ſuch ſort, that being full, that ſame extenſion cannot poſſibly receive more; but if we aſk, of what the extenſion of the atom is full, that is, in other words, what makes the atom ſolid and makes it matter; this the modern philoſophy does not teach. It is ſaid to be very hard, ſo hard as never to be broken by any force; but what makes it hard, or prevents its breaking is

not

not determined. It is faid to be inactive, and this inactivity is faid to confift in a tendency to perfevere in either ftate, of reft, or uniform rectilineal motion, which it at any time may happen to be in; that is, it may be either a *quiefcent* inactivity, or a MOTIVE inactivity, as accident determines; and this inactivity, thus explained, is fuppofed to belong to the whole atom collectively, and to every part diftinctly, if parts are diftinguifhed in it.

This idea of a fulnefs which refifts, becaufe, being full, the extenfion fo filled can admit of no increafe of fubftance, appears to be founded upon an erroneous notion of fulnefs; the fulnefs of any extenfion does not imply the exclufion of new fubftance from entering the fame place already full. For fulnefs does not imply refiftence, becaufe refiftence is action, but fulnefs is not action. If a place be full, no part of the place is without the filling fubftance, or, the filling fubftance is prefent in every part of a place which it fills. But a fubftance which *fills* a given extenfion may be more or lefs denfe; for if a fubftance can exift lefs denfe than another fubftance, there can be no reafon why this fubftance, of lefs denfity, fhould not be prefent *in every part* of a given extenfion; but if prefent in every part, there is no void or interftice, and the extenfion

fion is *filled* with the lefs denfe fubftance. Then fince, alfo, a more denfe fubftance may exift, this may alfo fill an equal extenfion, and the given extenfion filled with a fubftance more denfe, contains more, than when filled with a lefs denfe fubftance; for although in both inftances every part of each extenfion has the fubftance prefent, yet every equal part has *lefs* of the lefs denfe fubftance, confequently, that whole extenfion contains *lefs*, which contains the lefs denfe fubftance.

H Hence an extenfion *full* of a fubftance, may receive *more*, by the addition increafing the denfity of the filling fubftance.

I But that we may fee how errors, in like manner with truths, ferve mutually to each other's fupport, we muft obferve that the term *denfity* has been perverted from its fimple and familiar acceptation, and adapted to the purpofes of a theory, at the expence of reafon and nature. Denfity, in its common and juft fignification, imports that relation which arifes from a variation of the quantity of a fubftance in a given magnitude; thus if A, in a cubic inch, contains twice the quantity of B in the fame magnitude, A has double the denfity of B; but in the ufage of the modern philofophy, denfity is made to

fignify

signify the greater or less porosity of a body, and is in fine not at all different from porosity. But the better to disguise this fallacy, the language employed on this subject avails itself of the imperfection of our senses, and substitutes apparent and false magnitude, for actual and true. The *true* measure of a fabric, or a machine, is the extent of its boundaries, because the spaces within arise from those distances of parts which properly belong to the object; but in measuring a body, the true magnitude is the pure extent of solid parts; for the internal vacuities, though they go to form a house, do not go to form matter; the solid parts only, and not the vacuities, are matter. The *common* measure of bodies comprehends a great quantity of invisible pore, that is a great extent of vacuity, and is, therefore, though the *apparent*, a false measure of the extent of solid matter. To obtain the true magnitude of a mass of matter, the extent of pore must be taken from the measure of the whole body, which is interfected by pores. Now if, of two bodies, A be more porous than B, philosophers will say A is less dense than B, because in equal magnitudes of each; *ex. gr.* a cubic inch, A having more pore will have less matter; and to have less matter in an equal magnitude, is to be less dense. But the truth is, that a cubic inch of A is not an equal magnitude of *matter* with a cubic inch of B; A having more void within its limits, has lefs *extent* of *matter* than B, and, therefore, ought to have lefs *quantity* to be equal in denfity. To obtain the true denfity, we muft have the *quantity* of matter in a known *extent* of *pure folid*, and only as this quantity varies, denfities are different.

K   Solidity, faid thus to confift in a fulnefs of a thing unconceived; to have a hardnefs, the manner or caufe of which is unknown; to have an inactivity participating of fuch different natures, as motion and reft, is a very obfcure idea; and we have fhewn that this idea cannot be conformable to nature and fact, which prefent us ob- obvioufly with a very different one.

L   We have fhewn that folidity confifts in the prefence of a fubftance which excludes other fubftances from the place it occupies, not by *filling* the place, but by *acting* on that which it excludes; that this action conftitutes refiftence, whether to being penetrated, or broken; that inactivity belongs only to any folid taken as a collective whole, but that all the parts, confidered feparately, poffefs activity; that the inactivity of which we fpeak, and which we reftrict to the collective whole, but deny to the component parts of a folid, is not to be underftood, like

the

the inactivity of the modern philofophy, as implying a *tendency* to perfevere in any prefent ftate, of motion, or of reft, but as a privation of all tendency, and a reft confequent of that privation, except inafmuch as extraneous activities are fuperinduced.

This being our idea of folidity, we conceive of an atom as the leaft folid whole, as the leaft extenfion in which an active being is prefent under the conditions above defcribed; that is, acting fo as to exclude other beings, fo as to preferve its own figure and dimenfions from change, and by thefe two actions, preferving itfelf folid, and hard, and fo as by the equality of contrary actions to render the whole inactive.

An atom being confidered as the leaft folid whole. If feveral atoms ftick together by any means, they will make a larger mafs; but in this adhefion it is eafy to conceive that without a peculiarity of figure and arrangement the atoms will not come in contact in all points mutually; fo that in a mafs formed of feveral atoms there may be interftices between them, in which interftices there is nothing active, and which therefore may be eafily penetrated, and thefe interftices are called pores.

Conformably

O   Conformably to the foregoing ideas, the following definitions may now be given.

P   SOLIDITY IS THE RESULT OF THOSE ACTIONS AMONG THE PARTS OF ANY WHOLE, WHEREBY THE UNITY OF THE WHOLE IS PRESERVED WITHIN ITSELF.

Several uncohering things may be united by an external band; this does not conftitute thefe one folid, it may be one bundle; but if feveral things cohere and have a unity preferved within themfelves, they become one folid.

Q   AN ATOM IS THE LEAST AND MOST SIMPLE SOLID.

R   Having thus brought into view the nature of an atom, in order to prepare us to inveftigate its formation, and this nature appearing to confift in activity and mutual actions of parts, we muft now alfo recall attention to what has been fhewn concerning activity and action.

S   By *activity* is meant whatever has a difpofition to motion; by *inactivity* whatever has no difpofition to motion. Bodies have been fhewn to be active in two forms, actual motion, and a meer tendency to motion, as in impulfe; but the ACTIVE SUBSTANCE has been fhewn to have

but

but one manner of activity, that is, by its motion, becaufe it is incapable of impulfe; and there can be no activity in a perfect reft, void of any tendency to motion; therefore the ACTIVE SUBSTANCE moves as much as it is active.

Therefore we may fpeak indifferently of the *motion*, or of the *activity* of the ACTIVE SUBSTANCE, as fignifying the fame thing; for it is *active* inafmuch as it *moves*, and it *moves* inafmuch as it is active, and the quantity of the *motion* is proportional always to the *activity*; and again, converfely, the quantity of *activity* is proportional to the *motion*.

All the foregoing things relative to folidity, atoms, and activity, being fuppofed, the following problem is required to be folved:

TO EXPLAIN THE FORMATION OF AN ATOM OUT OF THE IMMATERIAL ACTIVE SUBSTANCE.

Whatever portion of the ACTIVE SUBSTANCE is given to form an atom, the followings things are neceffary to be united in fuch portion of ACTIVE SUBSTANCE.

Firft,

z   First, it must in some respect continually move (o 106), for otherwise it would lose its nature and cease to be active.

A   Secondly, it must also, in some other respect, be at rest, for otherwise it could not form an inactive atom.

B   Thirdly, it must preserve a unity within itself, by acting so as to resist whatever tends to destroy its unity (P, Q 164).

C   The two first requisites can only be united in a rotation of the portion of ACTIVE SUBSTANCE about a centre or axis at rest. By such a motion *all the parts* of the substance successively occupy different places in the orbit of rotation, and therefore move; but the centre round which they revolve being at rest, the *whole portion* is also at rest; and thus the portion is at once moving and quiescent as is required.

D   Nor will any other motion than this answer to the requisites given; for if all the parts move, as by the first requisite, and do not revolve about a quiescent centre, the centre, not being quiescent, must move together with all the parts, and the whole will move, not only in its parts, but as a collective whole, and cannot therefore be

in

in some respect, at rest, as the second requisite makes necessary.

The same kind of motion will also fulfil the terms of the third requisite. For a substance having a revolving motion around its own centre, preserves its unity by reason of all the parts preserving the same relation to the centre; and further, a motion of the ACTIVE SUBSTANCE about a centre or axis, will be an activity in the same orbit (T 165), which will act upon and resist whatever shall interfere to oppose this activity, or destroy the unity of the sphere, by diverting the course of the revolving motions.

The activity or motion of a portion of ACTIVE SUBSTANCE about a centre, will, therefore, give solidity to such portion, for it will give it unity and resistence, and in a manner tie together all the parts, forming them into one mass about their common centre; for they move or are active not *towards* the centre, in which case they would be lost in non-extension; nor *from* the centre, where they would dissipate in boundless space; but *about* the centre, preserving the same limits of extension; and being in this way active, they in this way resist any other activity opposed to them, that is, they resist any action

which

which tends to penetrate or divide this fphere of revolving activity.

G   Therefore, fince any portion of ACTIVE SUB-STANCE does, by revolving about a centre, become an united, refifting, and quiefcent whole. The fmalleft portions of the ACTIVE SUB-STANCE which have fuch motions, will become atoms, or make the fmalleft portions of matter (ℚ 164).

H   Since, therefore, an atom is actually formed out of the ACTIVE SUBSTANCE by fome certain modification of its activity; and fince the modification above defcribed, or a revolving motion about an axis will produce an atom, and no other modification whatever will ferve this purpofe; it is a juft inference from thefe, that the ACTIVE SUBSTANCE does in certain cafes and in fmall portions, affume revolving motions, each portion about its proper centre; and that thefe portions do by thefe modifications of their motions become corporeal atoms.

I   For the admiffion of any fact implies, alfo, the admiffion as a fact, of whatever is neceffary to the former; and *that* is neceffary to any fact, which will alone ferve to explain it: now refiftence can be explained only by action, and

action

action can be explained by the motion of an active being alone; and the refiftence on all fides of an extenfion, fuch as a folid body, can be explained only by an action on all fides; and an action on all fides can be explained only by a motion of an active being circumfcribing on all fides, the extenfion which refifts on all fides; and the quiefcence of this refifting extenfion, cannot be explained but by this circumfcribing motion being equal and contrary on the oppofite points of its motion; that is, in effect, by revolving in a fphere, about a centre, which centre, fo far as its own revolving motions are concerned, is at reft. Such a modification of the ACTIVE SUBSTANCE being, therefore, neceffary to the exiftence of an atom, or of any folid mafs, muft be admitted as matter of fact upon inference, becaufe neceffary to a fact afcertained upon fenfible teftimony.

Nor do I conceive this proof to be in its nature at all imperfect, or to fall fhort of demonftration: but if any one will refufe it, it will be neceffary for him to fhow, either, that the explanation offered is not fufficient, or that fome other explanation may ferve equally well.

We may now alfo explain what at firft may have appeared a paradox; how the ACTIVE SUBSTANCE,

SUBSTANCE, retaining its own nature and effential properties, continuing immaterial, unfolid, and active, puts on at the fame time the form of matter, and becomes material, folid, and inert (H 150). A fphere of revolving ACTIVE SUBSTANCE, as it revolves continually about a centre, and as parts of the fubftance, are confidered as fucceffively paffing through every point in the orbit; confidered thus, in its parts, and in its motions, it is ACTIVE SUBSTANCE, immaterial, and unfolid; but the whole fphere, confidered unitically, collectively, and as quiefcent, is, in this point of view, a folid atom, material, and inert.

M Whether every reader will poffefs himfelf clearly and fully of my idea I know not : I have endeavoured as much as poffible to diveft it of the obfcurity apt to attend abftracted reafonings; and I believe, but little difficulty will arife to any intelligent mind, except fuch, who, in their affociations of ideas, and acceptation of terms, have already, for a length of time, been habituated to a ufage foreign to that herein adopted.

N It is proper here to notice, that the fimple caufe, a revolving motion round a centre, produces the twofold effect, refiftence and reft, or folidity and inertia, agreeable to the order of nature

nature in other cafes, which is always to derive more complicated effects from more fimple caufes; but the modern philofophy, in its principles, violates this order, inafmuch as it requires two caufes, VIS IMPRESSA and VIS INERTIÆ, to motion, a fimple effect.

The relation between the ACTIVE SUBSTANCE O which forms an atom, and the atom, may be aptly refembled to the relation between a line and a circle.

The motion of a point generates a line; a line P may therefore reprefent the motion by which it is generated. If the motion of the point defcribes a part of the periphery of a circle, the periphery being completed, the moveable point has returned to the fame fituation from which it departed. The line being fuppofed to reprefent the motion by which it was formed, does this by the quantity of the diftance to which its extremity has elongated from the point of departure.

The line, while it is defcribing half the peri- Q phery, continues to increafe its diftance, and reprefents a progreffive and increafing motion. While it is defcribing the remaining femicircle, it is decreafing its diftance, till at length the moveable point coincides again with the point of departure.

R   Every portion of the periphery reprefents a motion, equal to the diftance between the two extremities of that portion; but the whole periphery reprefents no motion.

S   The periphery of the circle, therefore, confidered in portions, exhibits a fet of motions continually progreffive, and every motion of fome quantity or value, but whofe collective fum or refult is nothing.

T   And the reafon of this is obvious; for the motion in any part of the periphery, is, in a direction, contrary to that of the motion in the oppofite part of the periphery; and the fum of the motions, in any half of the circumference, are contrary to the fum of the motions in the oppofite half; confequently, fince the motions in the whole circumference are contrary and equal, their refult is nothing.

U   A circle, therefore, is a figure defcribed by motion, but which does not move; or, if the circle continually revolves about its own centre, the centre being at reft, and the periphery revolving within itfelf, then every part of the periphery,

periphery, and of the circle, moves, while the whole circle is at reſt.

A wheel turned round upon its axis, is an elucidation of all theſe obſervations, and an example of the union, in the ſame object, of activity and inertiæ, motion and reſt, to which we have traced the formation and exiſtence of every material atom.

If we employ a geometrical figure, as a circle, to explain any thing here advanced, we mean not, however, to offer this as a complete illuſtration of any phyſical problem; and it is the more important to obviate the danger of expecting more, in phyſics, from geometry, than geometry can effect, becauſe there may be reaſon to apprehend that the modern philoſophy is not entirely exempted from this error.

Geometry treats only of figure and magnitude, abſtracted from phyſical power, actually exerted. As a power, exiſting, and producing its effect, differs from the meer contemplation of that power in the mind, and the rationale of its quantity, ſo do phyſics differ from geometry; that is, as the havoc and devaſtation of a cannonade differs from the parabolic curve lines, in the books of gunnery, which, lying peaceably on their ſhelf, repreſent the courſes of projectiles.

Mathematics

A   Mathematics exift in the abftract reafonings of the mind; phyfics concern the phenomena of nature, the motions, buftle, and bufinefs of the world.

B   The ACTIVE SUBSTANCE is not an abftract idea, but a phyfical power moving and acting. Its figure and dimenfions may be fubjects of mathematical reafoning, but this reafoning can prove nothing concerning the exiftence of the phyfical power; that muft be proved in another manner.

C   It is an error into which numbers fall, to imagine the modern fyftem is either founded upon, or receives any fupport from mathematical proofs. The fcience of mathematics being confined to abftract ideas, can never afford any proofs concerning actual exiftence or efficient power; and he who fuppofes that the powers which actuate the planets and other natural bodies, are, or can be, determined by geometry, however fubtile or fublime, is as much deceived as he would be, were he to hope to afcertain the fecret proceffes of nature in the formation of the mineral, vegetable, and animal worlds, by reafoning upon the properties of a cycloid; would he develop the intrigues of courts from the nature of a fcalenus triangle; or direct the policy of
ftates,

states, by deductions drawn from the equality between a whole and all its parts. In fine, the error is no less than endeavouring to explain the œconomy of a *living* world by the science of *dead* quantity.

Should it therefore appear that the *mathematical* reasonings applied to the modern philosophy were in every part unimpeachable, yet it does not go the length of establishing a single fact; and the *physical* principles, incapable of receiving any such auxiliary support, must stand or fall by their own intrinsic merits or demerits.

And those who bestow such lavish encomiums on the invention which introduced geometry to physics, making thereby a new and mixt mathematics, should be careful to distinguish well, between *proofs* of actual existences as principles in nature, and the application of ingenious reasoning to *hypotheses assumed as data*.

Having seen the way in which an atom may be produced, we come to the second problem, viz.

TO EXPLAIN THE MANNER OF THE ATTACHMENT AMONG ATOMS.

The

11   The queſtion, how the parts of matter cohere, ſeems to have been the ultimatum of philoſophical reſearch. Men aſſumed the exiſtence of primary parts, and beginning with theſe, they endeavoured to ſolve the problem of the manner of their union.

1   But here, alſo, we find ourſelves enveloped in an obſcurity, the offspring of our own errors. For imagining all matter to confiſt of theſe primary atoms, and theſe atoms being conceived to be exceeding hard, impenetrable, and inert, two difficulties aroſe from this ſuppoſed ſtate of things.

K   *Firſt*, Not having inquired how the atoms were formed, or became hard, or exiſted, or how the parts of án atom were united; the light which the knowledge of this would have reflected upon the coheſion of the ſame atoms was wanting: for it has been ſhown (o 85, &c.) and it is without doubt, that we ought to conſider the union of two atoms, as a ſimilar effect with the union of the two halves, or any two parts of one, and as proceeding from a ſimilar cauſe; conſequently, the formation of an atom, had it been known, would have been a preliminary diſcovery, leading naturally to the queſtion of the union of two atoms.

Secondly,

Secondly, By assuming certain false data respecting the nature of atoms, men not only were deprived of an help which more just information would have afforded them, but became embarrassed with positive and insuperable difficulties, which arose not from the nature of the object investigated, but from their own erroneous data.

And these difficulties were two; *first*, from the extreme hardness and impenetrability ascribed to atoms, no substance could be supposed to hold together two or more atoms by mutually piercing each, as bodies are attached by cements or intermedia, nothing could, in this way, lay hold of an atom, to bind it to another. *Secondly*, all matter being thought to consist of like atoms, there was not any substance in nature which might serve as a bond of union between them: we were to conceive of no substances, but little unyielding masses, whose hard smooth surfaces might touch and easily slide over each other, but could in no conceivable way remain fixed together.

I shall not enter largely into the various opinions to which this inquiry and these difficulties have given rise. It has been supposed that the atoms were of an hooked form, and that these hooks,

hooks, by entangling each other, held the atoms together; the only intelligible fuppofition, as it appears to me, of which the notion of impenetrable atoms will admit.

More modern philofophers have rejected this fuppofition [a]; but, I think, neither on fair grounds, nor to fubftitute a better in its ftead: they reject it on the ground that it is begging the queftion; for, fay they, How do the parts of the hooks cohere? This fact being *fuppofed,* without explanation, cohefion is affumed to explain cohefion. But the charge which the moderns bring againft the notion of hooked atoms, does not hold good againft the authors of this opinion only, but againft all who maintain the doctrine of impenetrable atoms: for, whether the atoms be hooked, or fpherical, or of whatever form they be, they equally fuppofe

---

[a] Sir ISAAC NEWTON fays, " The particles of all hard homogeneous bodies, which touch one another, cohere with a great force; to account for which, fome philofophers have recourfe to a kind of hooked atoms, which, in effect, is nothing elfe but to beg the queftion. Others imagine that the particles of bodies are connected by reft, *i.e.* in effect, by nothing at all; and others, by confpiring motions, *i.e.* by a relative reft among themfelves. For myfelf, it rather appears to me, that the particles of bodies *cohere* by an *attractive force,* whereby they tend mutually to each other."

a power

a power by which their own parts cohere. Since, therefore, both parties alike suppose hard atoms, they may as well suppose them of an hooked, as of any other form.

Those philosophers who reject the explanation of cohesion by hooked particles, seem not to have been much more succesful in their own endeavours; they imagine that the particles of bodies cohere by an attractive force: but this, in fact, is no explanation at all; and, if it were, it is, on several accounts, liable to exception. It seems inconsistent to assign a *force* of any kind, as a cause of the cohesion of particles to each other, when the cohesion between the parts of any one particle, which is admitted to be vastly stronger, subsists without any force supposed. What is the origin and the seat of this force, it may be asked? How is it so connected with the inert particles, as to make them tend forcibly towards each other? How does it act: by impulse, or in what other way? Does it lay hold on contiguous parts, or encompass the entire atoms? Does it penetrate them, or act on their surfaces? What is its nature? Is it a substance, or a quality? Does it move, or remain at rest? In fine, what is an attractive force, its nature, origin, seat, manner of operating, and how is its existence proved? To all this it is no answer to fay, it is a force which acts upon all bodies univerfally. This is only rendering the pofition more general, but not more clear or more certain; for ftill the queftions recur of this univerfal force, in like manner as we have propofed them concerning the fuppofed attractive force of atoms. But further, as I faid, without going into thefe queftions, the explanation, admitted as it is offered, is, in fact, no explanation at all; for that can be no explanation which merely gives a new term, without affording a new idea relative to the difficulty which is to be explained. Now, the term cohefion is employed to fignify the fact of two bodies fticking together; attraction is ufed to fignify another fact, namely, the *tendency* of bodies, not in contact, to approach together; fo that if the tendency to approximate be called attraction, the fame tendency after contact may be called cohefion; and thus, attraction and cohefion fignify only different circumftances of the fame fact or tendency. Now, it certainly would not be thought to explain any thing if it were faid that the parts of bodies cohered by a cohering force: neither, therefore, can it explain any more, to fay they cohere by an attractive force; fince attraction, if it fignifies any thing different from cohefion, fignifies a prior fact, *i. e.* a tendency to approach before contact: but if the tendency towards each other,

in

in contact, called cohesion, is difficult to explain, is not the tendency towards each other, while at a *diftance*, equally or more difficult? So that the attractive force, being as obscure a thing as the cohering, it cannot elucidate the latter; and further, the attraction, if not the same with cohesion, differs only by including a prior fact; which, therefore, has no relation to the subsequent one, cohesion, required to be explained. Either, then, attraction is the same with cohesion, or it is a different fact, more obscure, and distinct from cohesion; and, in either case, it gives no new idea concerning cohesion, and, therefore, does not at all explain this fact, but is a new term, conveying either no new idea, or, what is worse, a foreign one, more obscure than that it is employed to explain.

From these views of the state of the inquiry, I think the conclusions are warranted, that the opinion of hooked atoms is unsatisfactory and hypothetical; and that the doctrine of attraction is still less admissible, because, at the same time that it is more unsatisfactory, it is less intelligible.

Respecting the union of masses, originally formed, and existing distinct, the following positions must be admitted:

Two

s   Two atoms are held together, or united by some force, power, or action, because there is a resistence to their separation.

t   The uniting power of atoms must belong to its proper substance, or there must be, together with an uniting power, an uniting substance.

u   The substance which serves to unite atoms, must have an union of its own parts; for, were it not united in itself, it could not impart union to any other thing; did not its parts cohere, they could not be the medium of cohesion to other parts.

x   Being a power or action, that which unites atoms is the ACTIVE SUBSTANCE, because nothing can act but what is active; and having an union of its own parts, it must be the ACTIVE SUBSTANCE, already constituted an atom; for an atom is the most simple united whole, formed out of the ACTIVE SUBSTANCE, as above described (g 168). Nothing more simple than an atom can exist, to unite other atoms; and nothing more compound can be admitted for this purpose.

y   Therefore, the atoms are immediately united to each other, without any medium, and without any other power than their own constituent power

or

or innate activity, whereby themfelves exift, as united and refifting wholes.

Two fubftances exifting diftinct in themfelves cannot become united while at a diftance from each other, nor by meer approximation and contact alone; for an union implies the deftruction of that diftinct exiftence which the two fubftances are fuppofed to have, prior to their union: but neither diftance nor contact deftroys that diftinct exiftence; the fubftances ftill remain diftinct, are two and not one.

In order, therefore, that two atoms fhould be immediately knit and held together of themfelves, and without any intermediate or foreign action (Y 182), there muft be fomething more than a touching of their furfaces, that is, there muft be an union or an intermixture, mutually, of part of the fubftance of each atom.

I fay there muft be an union of *part* of the fubftance mutually of each atom; for if the whole of each atom were mutually united, they would not be two atoms in union, but one atom: but part of each atom being united, reciprocally, there is then a continuity, an union of the fubftance of the two atoms, in fuch fort, that the two atoms can neither be faid to be any

longer

longer wholly diſtinct, nor are they become entirely one, but they are two atoms, whoſe ſeparate exiſtence is partially deſtroyed, and while they retain ſome ſort of diſtinctneſs, they alſo partake of an union; they are two atoms in one ſubſtance or continuous maſs. If two balls of wax are warmed at their ſurfaces and ſtick together, ſo that the figure of the two balls in part remains, but is deſtroyed at their junction, giving thus the idea of two balls, forming one piece of wax, this will be illuſtrative of the union, above explained, between two atoms.

c   This ſort of union, by a mutual penetration of parts with parts of atoms, we have here ſhewn to be neceſſary to any coheſion or attachment, whereby larger maſſes can be formed from the firſt atoms; and this neceſſity appears, and we have inſiſted upon it, independent of any theory reſpecting the nature of atoms, as admitting of ſuch mutual penetration, or not.

D.   The inert and impenetrable atoms of the modern philoſophy, certainly militate againſt any ſuch mode of coheſion; and it is as certain no other mode can be conceived, but either the hypotheſis of hooks, or the admiſſion of a new agent, contrived on purpoſe for this end; but any other agent is wholly unknown, and the idea

of

of hooks hardly deferves a ferious refutal, when we-confider the diverfity of forms and actions, among the minute parts of nature.

Our theory of atoms, comports with this mode of cohefion, which, independent of that theory, we have fhown, muft actually have place.

Our explanation of cohefion requires the mutual penetration of the atoms; and atoms, according to the foregoing theory, are not impenetrable, but refift only, in a certain degree, proportional to the quantity of the revolving motion, *i. e.* the action which conftitutes the atom; alfo, in a certain fenfe, atoms do not mutually refift penetration. The refiftence of an atom, or of any folid, is only to any other folid; but to the ACTIVE SUBSTANCE, confidered as unfolid, a folid oppofes no refiftence (F 85). Now an atom, being the fmalleft entire folid, is folid only as a collective whole, and as a collective whole it both refifts and is refifted. But the revolving motion, which forms the atom, has been fhewn to be, in every part, taken by itfelf, a motive, unfolid, ACTIVE SUBSTANCE: the revolving fubftances of two atoms may, therefore, mutually penetrate, near their circumferences, and interchangeably unite; and this portion, intermediate and common to each atom,

atom, will be the bond of union, and the tie which attaches them to each other.

G The theory of any particular fact must be allowed to receive considerable strength by its analogy with other parts of nature ascertained by sensible testimony. And since it is certain that all our knowledge of external things is attained through the medium of our senses, we can form no other ideas of beings and operations which are too subtile for the notice of our bodily organs, but by conceiving of them in a manner similar to actual phenomena. It is thus the hypothesis of hooked atoms, whatever objections it may be liable to, has some analogies, and is intelligible, because it corresponds with some sensible appearances.

H But the theory above laid down, has a far more extensive analogy, besides that it affords an adequate ground of application to the *varieties* of nature, and, above all, is not hypothetical, but deduced from phenomena. We see few, except artificial bodies, of an hooked figure, and attached by their hooks; nature, therefore, affords but a slender analogy of this kind. But, almost universally, we see bodies united by a mutual incorporation and introsusception of each others substances: soft wax, clays, gums, resins, refins, the grafts of trees, and the reunion of living animal fubftances, are examples. The fame thing happens when *intermedia* are employed: hard fubftances, whofe parts mutually refift penetration, cannot, for that reafon, of themfelves remain attached; but the glue which cements two furfaces of wood, in its fluid ftate, while hot, penetrates either fubftance, and unites its active powers with the contiguous corpufcles of the wood, and becoming fixed, when cold, the pieces of wood, thus united, on either fide, to the hard glue, are, by its intervention, firmly held to each other. Our theory defcribes the union of the primitive corpufcles agreeable to thefe fenfible facts, and this redundant hiftory of analogies. I truft it will appear more dignified by its fimplicity; and its confanguinity with familiar facts will be deemed, at leaft, one characteriftic of its truth.

The cohefion of the minute parts comprehends a great diverfity, which appears in the ftructure and modification of bodies: not only the varieties of hardnefs and foftnefs, brittlenefs and malleability or tenacity, fixednefs and fluidity, elafticity, and inelafticity; but no doubt, alfo, all the fenfible qualities of bodies, as odours, colours, vapours, and the ftructure and organization of different fpecies, both of inanimate and animate beings. A perfect theory of cohesion ought to afford explanations of all these diversities; it ought, also, to throw a light upon that opposite fact called repulsion, in which the parts of bodies manifest a tendency to recede from each other: but it came first in order to give, what hitherto has been a *desideratum*, a theory of the general fact of cohesion, previous to the investigation of its varieties, and deviations.

K And we are ready to confess, that although we can discern, in the theory laid down, a copious field of variety, adequate to all the purposes required, we have as yet seen but obscurely the immediate applications to each particular case; and without, at this time, prosecuting more into its minutiæ, an inquiry, the satisfaction resulting from which we cannot hope will be complete, we shall content ourselves with having cleared the way, and at least, as we hope, set forward in the right road, leaving yet an ample scope for the industry of others in making further advances.

L I shall conclude this part with a remark that I have frequently had occasion to make, the importance and pertinence of which entitles it to notice here.

I have

I have thought it neceffary to difcourfe, *firſt*, on the general fact of cohefion, and leave its varieties to a future difcuffion, of which the general principles muft be the bafis. But we continually fee men purfue a contrary method; they feek explanations of certain varieties of general facts, or deviations from the ordinary courfe of things, without thinking it previoufly neceffary to inveftigate thofe general facts, about which they are alfo ignorant. I fhall adduce fome examples to illuftrate my meaning.

Philofophers have long, as I before faid, confidered the queftion how the particles of bodies cohere to each other, without having firft afcertained what fort of things thofe particles were. This is as if a man fhould contrive a means of faftening together two bodies, without learning what fort of bodies they were, whether folid or fluid, foft or hard, ftone or vapour. Men began to inquire the caufe of eclipfes, who yet had never thought of difcovering the reafon of the more regular appearances of the fun and moon. The laws and caufes of the variation of the compafs have been the fubject of numerous inveftigations, but how few have thought it previoufly neceffary to inveftigate the caufe of the general tendency of the needle towards the North and South; and how neceffary is this inquiry as a

bafis

basis for the former: for the cause of each variation from itself, ought to be sought in some variation of the general cause, by which the needle points at all, and prefers one direction to another. Historians have laboured to develop the sources of wars, rebellions, revolutions, and remarkable events, with much more assiduity than has been bestowed to get just notions of the nature of that influence by which a government exists, and flows peaceably in its accustomed channels. Men are ordinarily solicitous to know the nature and cause of diseases, of blemishes, of preternatural appearances in our bodies, but indifferent as to the proceedings of the healthy œconomy: we have an hundred dissertations on fever for one upon life; we are more desirous to know how we are disordered than how we exist, and expect to be able to determine why the pulse beats unusually fast, notwithstanding we are ignorant why it beats at all. The action of stimuli, and the irratibility of the living fibre, have been the subjects of many ingenious discussions; the regular and uniform action of the fibre, of much fewer. A voluntary muscle in action, is contracted in its length, hardened in its consistence, and coheres more strongly, so as to sustain much greater tension, without dilaceration, than the same muscle not in its state of voluntary action; these accidental

conditions

conditions have, at all times, attracted the notice of phyfiologifts; but, in order to account for them, few, if any, have thought it neceffary to inquire firft, the manner in which the mufcle had its natural confiftence and degree of cohefion; expecting vainly to account for an irregular and extraordinary circumftance of cohefion, appearing in a living mufcle, whereby it undergoes great and fudden variations, dependent on a complicated fyftem of parts and connected agencies, without underftanding the nature of cohefion in its moft fimple ftate, between two primary parts of matter. With as reafonable a profpect of fuccefs, might a man, ignorant of the moft familiar truths in mechanics, attempt to explain the movements of the moft elaborate and complicated machine.

Various are the fources of this folecifm in the method of purfuing knowledge; habit, indolence, emulation, and other caufes, contribute to its production. It is the natural difpofition of the infantile and the untutored mind, to be taken only with novelties, and to be inattentive to accuftomed appearances. Every object, indeed, in nature claims our notice, but we cannot attend to all; familiar ones we regard with indifference, becaufe they are familiar, and fly with avidity to new fcenes.
The

The defire to exercife our talents, and achieve fome new difcovery, fets us to work; but indolence prevents us from carrying our refearches deep, in order to gain a firm bafis; we fkim the furface, and content ourfelves with fpecious errors, or pompous trifles.

In practical affairs, men are compelled to act upon imperfect knowledge: a general muft fometimes direct his movements on intelligence which may deceive him: a court muft give fentence, fubject to a poffibility of error: a phyfician muft prefcribe, on the moft probable conjectures, in an equivocal difeafe. But in thofe fpeculations, wherein immediate practice is not concerned, and which are undertaken, at our leifure, for the advancement of theoretical knowledge, and the benefit of men in the actual concerns of life, at a future period; in thefe, there can be no good reafon why we fhould begin at the wrong end, and wafte our labour in unavailing purfuits of what we can never attain, without going through a number of preliminary fteps, which we injudicioufly neglect.

C H A P.

CHAP. VII.

Of the Constructive Analogy between the Primary Corpuscles, the Planets with their Systems, the Solar System, and the Fixed Stars.

ON the same grounds, and in like manner as we have shown the construction of the primary particles, we may evince a similar construction to obtain in the larger parts of nature.

It will not be useless here to recapitulate the arguments before employed, previous to their more extensive application, as now proposed.

From the sensible qualities of matter we deduced, by inference, its actual but secret laws of existence.

The qualities from which we deduced the formation of matter and its manner of existence, were its SOLIDITY and its INACTIVITY.

We confidered fenfible maffes of matter as compounded of folid parts and pores, that is, as not entirely folid, or wholly matter, but comprehending void fpaces between the parts of matter; and we concluded that if any mafs were divided wherever pores interrupt its continuity, we fhould at length arrive at folid parts, which contained no pore: thefe imporous parts, ultimate in this order of decompofition, we called primary in the productive order of nature.

To thefe little imperceptible original maffes, corpufcles, or atoms, we transfer our ideas, and regard the qualities of SOLIDITY and INACTIVITY as fubfifting in them; and it is neceffary, in order to attend diftinctly and purely to the idea of folidity, to confider it uninterruptedly in a mafs devoid of pore, and which is folid throughout.

From *folidity*, as a fenfible quality, we infer *activity* as its immediate caufe; and we infer it on the ground, that folidity confifts in refifting action: we call that folid, which, when we prefs, or pull, or any way endeavour to change its dimenfions or figure, oppofes refiftence to our endeavours. In a more limited fenfe, fluids are folid, inafmuch as they refift compreffion. Refiftence to action, we fay, is action; an action

becomes

becomes refiftence, by its being confidered as oppofed to any other action, but without action there is no refiftence; for to refift is to impede, obftruct, or prevent, all of which imply action.

Since, therefore, folidity implies activity, every part of a folid atom being folid, every part is active, and the fubftance of the atom being folid, it is alfo active.

We regard an atom as the leaft and the moft fimple folid. If an atom fhould be divided, the divided parts would contain the ACTIVE SUBSTANCE of the atom, but they would not be fmaller atoms, or folid maffes lefs than the atom; for if the divided portions of an atom were yet folid, an atom would not be the leaft folid, nor a primary part of matter.

It is, therefore, only the *whole atom* which can be faid to be material, and the parts, confidered feparately, are not material, becaufe the material nature refults only from the whole conftruction. A certain conftruction of parts conftitutes a watch, but the whole, only, is a watch, and not each feparate wheel or part.

This follows, alfo, from what precedes (II 195): we have, by referring folidity to activity, and

C c 2   making

making an atom active throughout, because throughout solid, assigned to matter two qualities, which seem incompatible; or more strictly we have assigned to it a quality, ACTIVITY, and, at the same time, a negation of that quality; since besides solidity, the other property we discover, by our senses, in matter, is INACTIVITY.

M   *Inactivity* is an obvious quality of body; *activity* is a quality not obvious, but inferred from solidity. If matter be at once *inactive* and *active*, it must be in different respects: the inactivity being the more obvious quality, must belong to the whole collectively, and the activity being that which only is inferred, must belong to the parts. But since it is to the parts of an atom, taken separately, we must allow activity, and to the whole, collectively, we must deny it, the parts which possess activity, must be a different kind of being from the whole, which possesses it not.

N   Having thus been led to a division or decomposition of a substance, viz. matter, which produces a different substance, we resemble this process of ratiocination to those chemical analyses, whereby, from species of certain obvious and sensible qualities, other species result, the qualities of which were before concealed.

While

While we call the entire atom matter, it being like all matter folid and inactive, we call its conftituent parts, or the fubftance of the atom, confidered as unfolded, divided, and no longer held together in unity in the atom, the ACTIVE SUBSTANCE, and pronounce it IMMATERIAL.

We next inquire in what manner we are to underftand the activity of the ACTIVE SUBSTANCE, or to what intelligible idea activity can be referred, and we find it to confift in motion; for activity muft either relate to motion or to reft; but it cannot relate to reft, which is nothing, and requires no producer; it relates therefore to motion. In bodies, from whence we gather, originally, our application of the term activity, we find two ftates, in both of which they are faid to be active; both of thefe have a reference to motion; one is when bodies actually move, the other is when they impel or *tend* to move: nothing is called active but what does one of thefe. But impulfe is the effect of folidity, or of the refiftence of matter actually exifting; the ACTIVE SUBSTANCE, therefore, not being matter, cannot impel or be impelled, or cannot tend to move without actually moving: therefore, this immaterial fubftance muft move, and on its motion its activity muft depend.

We

Q   We therefore, now, may exprefs the *activity* of the parts of the atom, or the ACTIVE SUBSTANCE by its *motion;* and the *inactivity* of the entire atom, by its *rest;* and we may fay the *parts* continually move, while the *whole* is at reft, as we before faid, the parts were *active*, while the whole was *inert*.

R   As the parts conftitute the whole, the ACTIVE SUBSTANCE conftitutes the atom; and the ACTIVE SUBSTANCE, by its own nature, conftitutes the nature of the atom. Now, the nature of the ACTIVE SUBSTANCE is motive; and by this its motive nature, it conftitutes the nature of the atom, folid and quiefcent.

S   But in order that motion of parts fhould conftitute reft of the whole, the moving parts muft have contrary motions, in equal quantities, from which will refult the quiefcence of the whole; and to do this, the moving parts muft revolve about a centre at reft, either in circles or in fome curvilinear orbits.

T   And fuch a motion, befides producing the quiefcence of the whole, whofe parts fo move, will alfo produce the folidity of that whole; that is the orbicular motion will be an activity, capable, in every part, of refifting other activities.

And

And that which is at reft, and refifts in every part to extraneous actions, is a folid.

The orbicular motion of the ACTIVE SUBSTANCE, that is, of an immaterial fubftance, will produce an atom, or a material being; and no other modification of motion will afford this refult. On this ground we infer, that the ACTIVE SUBSTANCE does actually form fuch orbicular motions, about quiefcent centres, and does, by this means, conftitute the primary parts of matter.

Such, in a general view, has been the method of reafoning employed, and the fteps by which we have arrived thus far: and we fhall now proceed to a more extenfive application of the fame.

The obvious and fenfible qualities of bodies, folidity and inertia, we transferred, in idea, from fenfible maffes, to minute parts imperceptible to fenfe; but we may, alfo, confider thefe qualities in the larger portions of the creation;— the earth and the other planets: again, we may *infer* the fame qualities of other, and larger fpheres of exiftence; fuch as the orbits in which the planets, primary and fecondary, revolve; the

whole

whole mundane fyftem; the fyftems of the fixed ftars; laftly, of the entire univerfe.

A   Each, and all of thefe, refpectively, we may pronounce to be, in a certain fenfe, folid and inert; that is, to partake of the nature of matter.

B   We muft recollect, that the inactivity of which we all along fpeak, or the quiefcence we attribute to material exiftences, is not an *actual*, but a *natural* reft. All bodies are moveable, or fubject to occafional activities; but thefe activities are confidered as accidental or extraneous, and not as at all to be regarded in confidering the nature of bodies. No atom has an actual reft, although we afcribe reft to them. We mean, that reft muft be the refult of their own conftitution and innate action; and if they are made to move by foreign caufes, we do not take thefe motions into our prefent confideration. So of all the things above enumerated, when we pronounce them to be quiefcent, we mean that they are fo, relative to any internal activities which belong to them, and conftitute them, and have no motions but what are induced from without, and fuch they actually have.

Reft,

Reſt, or inertia, being thus underſtood, and ſolidity, as already explained, to conſiſt in a reſiſtence about a circumſcribed extenſion, theſe may be predicated of every diſtinct maſs, ſphere, or extent of being, from the primary particles, to the whole univerſe.

That the earth is a ſolid body, we know by our touch; that the other planets and ſtars are ſo, we gather from their emitting or reflecting light to our viſual organs, and from other analogies. The motions of the planets are conſidered as induced by extraneous cauſes; and relative to the actions carrying on within themſelves, they are at reſt; theſe actions, proper to each body, being on the whole, equal in contrary directions. Theſe bodies, therefore, have the qualities belonging to all matter, ſolidity and inertia.

That the ſpheres, circumſcribed within the orbits of the ſeveral planets, are alſo ſolid and inert, and conſequently material, is leſs obvious, and ſuch language is leſs conſonant to our habits of ſpeech; but, I apprehend, it may be made to appear with equal certainty.

A ſolid is a circumſcribed being, which, from its own internal activities, reſiſts on all ſides.

fides. No prefcribed quantity of refiftence is neceffary, that may vary indefinitely. In order to folidity, a being need not refift as much as iron or wood or air; it may be far lefs denfe than thefe, and yet be folid, provided it refifts on all fides.

G The term folid is applied to a fpecies of being having a determinate character, that is, a circumfcribed refiftence; not to any fixed degrees of that refiftence. As there may be invifible folids, like air difcovered by the touch, fo there may be intangible ones, difcovered by inference from other phenomena. Tangibility belongs to that degree of refiftence, in folids, which is capable of producing a certain change on fuch folids as our organs: a lefs degree will fail of being manifeft to our touch; but philofophers are abundantly convinced, that one fenfe is not the criterion of exiftence, but fuch inferences as refult from the teftimony of all. It is the property of fire to give the feeling of heat, but not of all degrees of fire; for we are affured of the prefence of fire in bodies, which are cold to our fenfe. Light may be proved to exift, where the eye would announce a total darknefs.

H In like manner, matter has its degrees of intenfity. Water is lefs denfe than ftone, air

than

than water, vapour than air, light than vapour; and as light is intangible, excepting to the optic nerve, why may not matter ſtill leſs denſe than light exiſt, to us inviſible as well as intangible?

We ſhall not, therefore, be miſunderſtood, when we ſay the planetary ſpheres are material, that we think them groſs, like the matter around us, or that we adopt the ſolid orbs of the Ptolemaic ſyſtem, or the denſe ſubſtance of the vortices of Des Cartes. But we ſay the celeſtial orbits exhibit evident proofs of their being *circumſcribed extenſions, reſiſting on all ſides,* and inert; and of having, thus, the characteriſtics of matter, and being altogether ſimilar to ſenſible bodies, except a difference in degree.

We know from obſervation, that a ſphere of activity extends to ſome diſtance around our earth: and we ſhall conſider it as a *datum* that ſimilar ſpheres extend around all the planets, and the ſun; by which they interchange a mutual influence; and, alſo, about the fixed ſtars, although their agency is leſs apparent to us. To what limits theſe ſpheres of activity extend, about each celeſtial body, we do not here inquire.

L   The phenomena of falling bodies in different parts of the atmosphere surrounding our earth may serve to illustrate the distinction we wish to have clearly understood, between the part of an atom, and the whole. Since, on every part of the earth's surface, bodies fall, the agency, therefore, by which they fall, extends all round the earth. The whole of this sphere of agency we call a sphere of matter, less dense than our earth, and enveloping it; but if we speak of any separate part, we call it immaterial. Thus it is an *immaterial* substance and agency which causes a stone to fall; because this is done by a portion of the sphere: in each portion the sphere is active, in the whole inert; in *each part immaterial, as a whole, a material* but rare and subtle orb.

M   The unity and order of each distinct system, and of the entire universe, demonstrate a mutual resistence to each other, and, at the same time, a mutual attachment of these spheres of activity.

N   The earth and moon, in their revolution about the sun, carry along with them their respective spheres of activity; but neither penetrates, nor is penetrated by the other. The same may be said of Jupiter and Saturn, with their moons; and

and all the planets, with the fun, revolving about their common centre, preferve their diftances from that centre; and fince the very being of the univerfe depends upon every part preferving its proper diftance from every other, there can be no doubt but a powerful agency is prepared to *refift* any tendency, actual or poffible, to the falling together of the celeftial bodies.

It is equally effential to the prefent ftate of things, that the planets are not too widely fcattered afunder, as that they do not approximate too near; and that every part, planets, fun, and fixed ftars, neither encroach upon each others limits, nor any of them fly off, detached and loft from the fyftem, into boundlefs fpace : hence it is as evident that there exifts an *attaching* force to preferve the connection and order of thefe parts, as it is that there exifts a *propelling* power, to prevent their rufhing together into one common mafs.

Without an attaching power, the unity of the whole could not be preferved; without a refifting, the diftinctnefs of parts muft be deftroyed.

This two-fold power, which is neceffary to the exiftence of the larger planetary fyftem, we have before fhewn to be neceffary to the exiftence of
a fingle

a single atom; that is, a refiftance of the parts to a nearer approach, and to a more diftant recefs (E, &c. 29), and in this we perceive a perfect fimilitude between the nature of the general fyftem, and of the fmalleft atom.

R   If we regard any fingle fphere of activity, we may refemble it to a fingle atom. It is a complete exiftence in itfelf; a whole, formed of its own parts. Thus we may regard the fphere of the earth's activity, of the moon's, of Jupiter's, Saturn's, or the fun's.

S   If we regard any of thefe fpheres in their relation to other fpheres, we may refemble them to atoms in their relation as cohering to other atoms.

T   Thus the fpheres of the earth and moon cohere together; of Jupiter and his fatellites; of Saturn and his fatellites; and thefe may be confidered, again, as compound maffes, which, together with the fpheres of the other planets, all cohere with the fphere of the fun's activity, or are involved in it.

U   Again, the whole fphere of our folar fyftem *coheres* with contiguous fpheres of the neareft fixed ftars; and each of thefe immenfe orbs of activity

activity may be regarded as an *atom*, in relation to the whole univerfe; and the univerfe as a *quiefcent mafs* conftituted by the cohefion of thefe atoms, in a manner analogous to the formation of any fmall and more denfe body, by its component atoms.

Our habit of conceiving the planets to revolve in fpaces almoft or altogether void, and as *attracting* each other, at great diftance, through this void, may impede the familiar conception of thefe fpheres of activity, as extenfions of actual and ACTIVE SUBSTANCE diffufed around each planet: but we have, all along, affumed it as evident, that where there is *action*, there is *fubftantial exiftence*, and confequently that to whatever apparent diftance bodies manifeft that power called *attraction*, it is not truly a diftant action, but the action of a fubftance, prefent where it acts, and confequently extended about the central globe (faid to poffefs the attractive force), to the diftance of the body acted upon. The planets, therefore, cannot be faid to attract or act upon each other, at a diftance, or through void extenfions; their fpheres of attraction, as they are called, are fpheres of an ACTIVE SUBSTANCE, lefs denfe than the planet which occupies the centre of thofe fpheres; and the active fphere of each planet acts not on the *body* of another

x

another planet, but on the *active sphere* of that other planet, extended so as to interfect the former. This idea is not diffonant to the opinion of Newton, who exprefsly rejects the notion of bodies acting at a diftance, as being what no philofopher can entertain.

y   The analogy we have obferved between the feparate fpheres of activity, which encompafs each their proper central body or planet, and the primary atoms; and between the attachment of thefe fpheres, and the cohefion of thofe atoms, leads us neceffarily to a fimilar method of invefligation in each, and we have here, as in the inquiry concerning the formation of matter, two diftinct facts to difcourfe upon, in their proper order; *firft*, the formation of any fingle fphere of activity; *fecondly*, the mutual attachment of thofe fpheres.

z   And in each of thefe, the fimilitude of the facts leads precifely, by the fame fteps, to the fame conclufion. (See Ch. 6. P. 2.) Each fphere, that of the earth for example, in all its parts, feparately taken, is active; whether we regard this fphere as acting on that of the moon, to retain it within its proper limits, or in the more familiar facts of gravitation, near the furface of the earth, we fee it actively demonftrated;

but

but the equality of all thefe actions on oppofite fides, makes their refult nothing, and the whole fphere as an unit inactive: accordingly, philofophers are agreed, that the earth is not moved by the gravitating tendencies upon its furface, becaufe they are contrary and equal.

But this activity of parts, and thefe various **A** actions of gravitation, carrying on above the earth's furface, in a medium of ACTIVE SUBSTANCE, are produced by the *motion* of that fubftance (z 166), and as the activity *exifts* only in feparate parts, and is *nothing* in the collective whole, fo the motions, in like manner, muft be *motions of parts*, which refult in the *reft of the whole*.

And fuch a motion can be no other than a **B** revolving motion about a quiefcent centre: AND HENCE WE INFER THAT THE ACTIVE SPHERES WHICH ENCOMPASS, SEVERALLY, EACH SECONDARY PLANET, PRIMARY PLANET, THE SUN, AND EVERY FIXED STAR, ARE SPHERES REVOLVING ABOUT THEIR PROPER CENTRES, AND THAT BY THIS REVOLVING MOTION THEY EXIST, AND PERFORM THEIR ACTIONS.

Refpecting the cohefion of thefe fpheres, we **C** fay they cannot hold each other together by any

E e   diftant

diſtant action; neither by ſimply coming in contact at their ſurfaces; but it is neceſſary that parts of two attached ſpheres ſhould interſect each other, and that their ſubſtances at theſe interſecting parts ſhould intermix, and their actions be blended in ſuch ſort as to conſtitute theſe conjoined portions, common to either ſphere, the bond of attachment between both.

D   In the celeſtial machinery there are two ſpecies of *attachment* to be traced. The *firſt* is that of ſpheres interſecting each other, in portions, the remaining portions extending without each other; the *ſecond* ſpecies of attachment is when leſs ſpheres are *wholly contained* within greater.

E   The ſphere of our ſolar ſyſtem can only be connected to the ſpheres of the fixed ſtars by interſection of portions of the ſpheres; but the ſmaller ſpheres of the planets belonging to our ſyſtem, are contained, and wholly enveloped within that larger ſphere.

F   When any ſphere of activity involves another ſphere, in ſuch a manner that the involved ſphere is *ſingly* acted upon by the other, in which it is contained, it will neceſſarily follow that this contained ſphere will be carried about the centre of the other ſphere, by the revolving motion of

that

that other; or, that each sphere will be carried about a common centre by the conjoined revolving motions of each.

From hence the revolutions, both of the primary and secondary planets, ascertained by observations, might be demonstrated *a priori*, could their relative distances be ascertained, independent of the knowledge of their motions.

The moon and earth, by their vicinity, ought to revolve, by the mutual actions of their spheres of activity, about a centre common to both.

The same may be said of the other satellites, respecting their primaries.

All of these, primaries and secondaries, being involved within a larger and parent sphere, ought to be carried about a common centre, by that larger vortex.

But, at least, the foregoing principles serve to afford a cause, true, sufficient, and intelligible, of the revolutions of the celestial bodies.

When, in the *second* species of *attachment*, any sphere, instead of being singly acted upon by another sphere, is, at several parts, interfected by by several other surrounding spheres, and attached, in this manner to each, such a sphere cannot be carried about the centre of any other sphere, because it is attached to *several*, but must remain at rest.

N   Hence appears the reason why the sun and fixed stars are, relatively to each other, at rest, and form a quiescent mass, or an universe fixed and immovable.

O   For the multitudinous spheres of the fixed stars, surrounding each other on all sides, and each being, in various parts, intersected by its contiguous surrounding spheres, and held to each by the intersected portions, every sphere must, thus, be retained at rest.

P   And wherever the universe has its boundaries, the external spheres, although they cannot, like the internal ones, be surrounded by other spheres, may yet be each attached to several others, and being held by each, be prevented from obeying the revolving motion of either, and thus, also, retained at rest.

Q   And if our visible universe be a whole, unconnected with any other creation, and not forming a part of a still larger system of being, it ought,

by

by our principles, to poffefs ACTUAL REST, while all its component parts fubfift, and conftitute that whole, by their ACTUAL MOTIONS.

But if our vifible creation be only a part of a larger one, a link of a more extended chain, which embraces wider and remoter regions, than thofe from whence the fmalleft telefcopic ftar darts its rays, then by the fame principles, and by the general analogy of things, *our univerfe* muft have its *proper motion* in that LARGER SYSTEM, of which it is but a component part, and which other and unknown worlds combine to frame.

END OF THE SECOND PART.

PART III.

A further Investigation of the Nature and Laws of the ACTIVE *or* ELEMENTARY SUBSTANCE.

CHAP. I.

Origin of Physical Power, exemplified in Elastic Bodies, Inflammable Bodies, Chemical Operations, Living Processes, and Gravity.

THERE is a power of volition, and a power A which, so far as observation can go, appears to be independent of volition; this latter is called physical power.

Activity, and consequently power [a], has been B found to have its origin in motion.

[a] See Definitions.

But

C   But it is phyfical power or activity, alone, which we mean to confider as derived from motion. The activity, or power of mind, we do not, in this place, take into confideration.

D   If power proceed from, or rather belong to motion, whence does the motion, in all cafes, proceed?

E   An impelling body, being in motion before it impelled, may be conceived to be a vehicle of fomething motive, which it can impart in the impulfe; but when no motion is feen to *precede* an action or exertion of power, only to *follow* it, whence can it be fuppofed the motion proceeds? Whence, the motion of falling, generated in a body, when left without a fupport? Whence that, in an elaftic fubftance, left to itfelf, after compreffure? Whence that, proceeding from inflammable bodies, as gunpowder? Whence the inteftine motions of chemical changes, thofe of vegetation, or of animal life?

F   It appears from what has been fhewn to be the conftitution of matter, that all bodies contain within them a fource of motion, though themfelves be at reft.

For

For the primary component parts, or atoms, G are so many little spheres, revolving about their proper centres; and thus the whole mass of every body is made up of motion, exists by motion, and is full of motion; it is, therefore, an obvious source, out of which motion may proceed.

But bodies neither are moved by their own H internal motions, nor immediately generate other motions except placed in particular circumstances, not natural: an elastic body produces no motion unless it be first compressed; nor inflammable bodies, unless fire be applied to them.

Some accessary circumstances are, therefore, I necessary in order that bodies should *give out* a portion of their internal motion, which, in their natural and ordinary state, is destined and employed for the maintenance of the body, and which circulates and abides within it.

The bending of steel, or of any elastic body, K puts it in a preternatural state. The bended body presses on whatever detains it, and by pressing, continually *pours out* a current from within it (R 127); and when the body is no longer forcibly detained in a bended form, it restores itself, that is, its parts begin to move, and regain their natural state, after which the body exerts no impulse,

impulse, nor *gives out* any motion. Agreebly to this, most bodies are found to diminish of their elasticity, after a time, by remaining bended; the internal structure is changed, or its force weakened, by the loss of so much motion, destined for its support.

L If A. M. fig. 1. be an elastic body not bended, and p. p. p. be primary atoms of the mass, intersecting each other, so as to cohere; these atoms will have each its proper revolution about its own centre, returning always into the same orbit.

M . Let A. M. fig. 2. be the same body bended, the atoms p. p. p. will then be flattened, and the atoms q. q. q. will be elongated. This change being preternatural, every atom will tend to reassume its proper orbit, and this tendency will produce the spring of the body to unbend itself; and as it presses outwards, part of the revolving matter of each may penetrate the neighbouring atoms, as represented by the dotted lines; and thus getting out of its due circulation, and being in a manner *let loose*; such part of the revolving matter of the atoms may pervade the mass, and become a current of ACTIVE SUBSTANCE flowing out, at either end, of the bended elastic body.

Inflammable

Inflammable bodies, by the acceſſion of fire, a ſubſtance highly active, acquire alſo that ſame active ſtate; but by thus in a manner letting looſe their activity, the ſtructure is damaged and deſtroyed, the revolving ſpheres which formed the atoms of the body are unfolded, at leaſt in part, and the ACTIVE SUBSTANCE of the body aſſumes a new form, more near to its priſtine ſtate. Accordingly, we find inflammable bodies wholly deſtroyed by combuſtion, and ſee them diſſipate in the form of more ſubtle elements.

In chemical operations, and the proceſſes of living bodies, the inteſtine motions proceed from and ſerve to prove the reality of the conſtituent revolving motions, by which the bodies exiſt. As each atom exiſts and holds to other atoms by its revolving motion, it is eaſy to conceive that a great diverſity may obtain in their arrangement, and that they may have various relations among each other, and various affinities, whereby parts in union may ſecede, and new unions form, and different orders of parts compound and decompound; and hence ſolutions, precipitations, chryſtallizations, fermentations, evaporations, ſublimations; hence digeſtion, ſecretion, circulation, aſſimulation, and the various animal functions.

P   The revolving motions by which matter exists in *any form*, is a GENERAL BASIS for the internal powers, which, operating by *particular* and unknown laws, produce the various *changes of form*. The revolving motions afford a *physical power*, actually present in all bodies, and adequate to the productions of all their changes; it is only concerning its particular *laws of operating*, to produce those changes, that we are yet ignorant (K 188).

Q   There is a physical power which does not reside in bodies, usually denominated *gravity*; it is that whereby bodies fall, and are heavy. This power surrounds the earth, and it is probable, all the planets, to great distances. And since all physical power consists in motion, motion must therefore surround the earth, and planets, to whatever distance gravity extends about them; such are the spheres of activity above described (K 203).

R   But this is a motion differently circumstanced, in respect to us, from those, which, by the revolutions of primary atoms, form bodies.

S   An atom is a small sphere, the resistence of which we discover from *without*, and on all sides of the sphere. The action which surrounds our earth,

earth, is a larger fphere, *within* which we ourfelves are contained; and we can only, at any one time, and place, experience the refiftence in one certain direction, *i.e.* towards the centre.

Again, an atom is a fphere of greater denfity, which affords a more ftrong refiftence; and the fphere which furrounds the earth, a fphere of lefs denfity, refifting lefs powerfully. An atom in an extenfion which is too fmall to be perceived, refifts too powerfully for our ftrength to overcome. The action of gravity, in a much larger extent, we can overcome, and in very fmall extenfions it is not perceptible: we *feel* no weight in a very fmall body, which yet may be fo hard, that a man's ftrength cannot break it.

Thus we find that nature contains within itfelf all the powers neceffary to her œconomy; and if it be true that power is neceffary to a continued exiftence of our being, in any form, that conftituent power is to be regarded as the agent in thofe changes of form, which all things, fuccefsively, affume.

CHAP.

CHAP. II.

Origin of Motion.

A IF power originates from motion, it seems next in order to inquire from whence motion arises.

B We easily gather, from observation, that bodies do not begin to move from any cause within themselves, and thence we are led to seek an external cause.

C We see that bodies are moved by the impulse of other bodies, and hence we know impulse to be an external cause of motion; an impelling body did not acquire motion from within itself, but from an external cause; and if this external cause were another impulse, the cause of the motion of that last impelling body is to be sought: and as successive impulses are not infinite, there is always a first impulse, in every series, which did not derive motion from impulse, and consequently

quently it muſt have derived it from ſome other external cauſe.

That other external cauſe of motion, beſides impulſe, we are to inquire.

This cauſe cannot be body, ſince body is of itſelf quieſcent (B 222).

For this reaſon, obſervation cannot, in this reſearch, afford us any immediate aid, ſince only bodies are the ſubjects of our obſervation.

We muſt, therefore, proſecute the inquiry by the exerciſe of our intellect alone; obſervation can only aſſiſt us in a ſecondary way, by affording analogies, and other materials, to reaſon upon.

We are firſt to ſeek ſimple and intelligible cauſes, before we admit obſcure and myſtical ones.

We are to prefer a cauſe analogous to other known cauſes, before one which is different from the viſible order of nature.

The moſt natural and obvious analogy by which, in our preſent inquiry, we are to reaſon from

from facts *observed*, to the discovery of an *invisible* cause, is that of *impulse*.

L   Impulse is the only visible cause of motion; an invisible cause, which is in any way allied to impulse, must be preferred before one wholly unconnected with it.

M   It is from motion bodies derive their *impelling* power, and become, themselves, causes of motion.

N   The cause we now seek is not body (D 223), and therefore, in respect to its *substance*, has no apparent alliance with *impulse*, which is performed by body: but since there are only two things necessary to an impelling power, viz. body, and the motion of the body; and the cause we now seek ought to be allied to impulse, and is not in its *substance* allied to it, as body, it can only be allied to it in its *mode*, that is, in its *motion*.

O   The invisible and original cause of motion ought, therefore, to be a substance, motive, but not corporeal.

P   Thus we arrive at the same principle, the ACTIVE SUBSTANCE, which to discover and understand, has been the main business of the preceding pages.

We

We are then led to conclude, that bodies Q which are moved, and not by impulse, are moved by a motive substance, not corporeal.

If we were next disposed to ask, how this in- R corporeal substance moves bodies, we are ready again to study the *analogy* of the *sensible* cause.

Bodies, in impulse, act by their presence con- S tiguous with the body moved, and not by contiguity alone, but by *pressure*, whereby the agent is disposed to penetrate the patient, which in soft bodies it does, but in hard is resisted.

So also the incorporeal agent ought to act by T its presence, contiguity, and *disposition* to penetrate the patient; but being incorporeal, and incapable of being *resisted* by hard bodies, it ought also actually *to penetrate* all bodies that it meets in its course; and whereas *bodies* act by their disposition to penetrate, *this* ought to act by its actual penetration.

From the motive nature of this substance, and U the perfect permeability of bodies to it, its successive *flow* through series of impelling bodies, and in all impulse, is easily proved, as may be seen above (U 128).

G g Hence,

x   Hence, from the analogy of impulse, we discover an invisible original cause of motion, which serves reciprocally to throw a new light upon impulse itself, from whose aid we were led to it.

y   For impulse, although it is a *visible* cause, is yet an *obscure* cause of motion. All that we see of impulse is the approach of the impelling body, its contact and its pressure on the patient; we see motion in the patient follow these appearances, but an obscurity still pervades the fact; for all appearances, in impulse, take place *without* the patient; motion, their effect, exists *within*; what, therefore, conjoins the *external cause* with the *internal effect?* Here the continuity of our ideas is interrupted, the chain is broken, a chasm presents itself, in which our minds are bewildered, without light or guide.

z   The continuity is restored, the link supplied, the chasm filled up, a light and a guide are found, when, in impulse, we trace the invisible flow of the pervading ACTIVE SUBSTANCE, conveying from the vehicle *without*, the impelling body, to the new receptacle *within*, the newly moved body, an actual and efficient cause.

The

The ACTIVE INCORPOREAL SUBSTANCE, by its motion, is the origin of motion in bodies; what then the origin of motion in this incorporeal substance is, we must next inquire.

We suppose that all beings, save the Supreme Cause, are the creatures and instruments of that cause.

Every such subordinate being must exist in a state of motion or of rest.

But that which exists in rest, can have no subordinate activity, cannot be an instrument of another; absolute rest is absolute death, the privation of all power, all faculty, all instrumentality; it is pure and abstract passivity, it can suffer agency, but exercise none[b], it can fill up no place in a chain of being, because incapable of any agency, even in the lowest degree subordinate.

[b] In strictness, absolute rest and inactivity cannot have existence, and therefore cannot suffer agency, any more than exercise it. Existence itself, implies activity; it is, therefore, only conditionally said, that which exists at rest (if any such being does exist), can suffer activity, that is, if it can be conceived to exist, it can be conceived to suffer.

E   If fubordinate beings were created from nonentity, would the Creator have produced an *inactive lump of exiflence*, as the immediate work of his hands ? Could death have been the offspring of the Parent of life ?

F   If inferior beings have exifled always, and if any fuch were always inert, from inertia, what activity could arife, what change, what fucceffion of exiftence, from the negation of all faculty and all power ?

G   If the firft origin of things, if Deity be an active caufe, fo, alfo, muft fecondary and inftrumental origins be. If inertia can be an *immediate* agency, it may be a *remote;* if inertia can *act* in *any* degree, it can act in *every one ;* either we muft exclude *abfolute inactivity* from exiftence, or allow of no other exiftence ; it can do nothing, or it can perform all.

H   Whatever, therefore, be the elementary fubftance from which creation affumed its prefent form, whether, an incorporeal uncohering being, or whether, as is the prevailing idea, in this day, certain primitive bodies, fmall and very hard ; this elementary being, whether produced from nothing, by the creating fiat, or exifting from eternity,

eternity, *cannot be inactive, cannot have existed originally in a state of rest.*

If the elements of things were naturally at rest, motion must be to them an unnatural state, to which they must be repugnant. Nor can it be conceived how that to which rest is natural, can exist in motion; or how the natural state of a being can be destroyed without the being losing its nature, and being itself destroyed.

Since, then, rest is not the natural state of the elements of matter, their natural state must be motion.

The motion, therefore, of the elements of nature, originates only with their existence, and is coeval with their birth. Motion is the primordial state of creation, which it cannot lose without ceasing to be.

Since then motion is a state original and essential, rest must be a condition derived, subsequent and accidental.

And since all things that exist as derived, subsequent and accidental, must have been produced by those which exist as original and essential; rest must be the production of motion.

There

o   There are two relations of motion which produce rest, equal concurring and equal contrary motions.

p   In equal concurring motions, all the parts of the conspiring motions are relatively to each other, at rest; but the whole are in motion.

q   In equal contrary motions, the parts which move contrariwise, move relatively to each other, but the whole is at rest.

r   The rest we experience on the earth's surface, in a carriage, or a vessel in motion, is of the former kind, a relative rest, among parts, of a whole in motion.

s   The rest of the solar system, or of the universe, is of the latter kind; the motions of all the parts result in the rest of the whole.

t   I have not, in discoursing on motion and rest, used the distinction generally adopted, of absolute and relative motion and rest; both because it did not serve my purpose, and because I am not satisfied of the propriety of those distinctions. I rather apprehend that all motion is necessarily relative, and all rest relative to motion; for motion being a change of place, and place being the

relation

relation between bodies compounded of diftance and direction, that is, a relative thing, motion is only the change of relation between bodies. Reft is the relation of equality and agreement in motions (P 230), or of their equality and contrariety (Q 230). In the one cafe, on *comparing* confpiring motions, their *difference* is found to be nothing, in the other, on *fumming up* contrary motions, the *refult* is a cypher.

If, not confidering the true nature of folid U matter, we fhould imagine it to be a pure inactive effence, void of motion, or tendency to motion, and impenetrable, it feems to me that no natural origin, or caufe of motion, could ever poffibly be affigned. For fuch a caufe muft either act by its own motion, or by fome other way. If it were fuppofed to act by its own motion, what being do we know that has motion in itfelf, or how fhould we reafonably fuppofe that one fort of being, matter, was made without motion, and another fort of being was made on purpofe to give it motion that was not natural to it. But if we fancied a being made motive on purpofe to move matter, we could not conceive how matter could be moved by it, or be moved, indeed, at all, if reft was its proper nature.

Matter

x   Matter being impenetrable, nothing could be communicated to it, to fuperadd a new nature, and nothing exifting without it could be conceived to produce fuch a change. If we fuppofe motion originated from fomething not motion, this muft be confidered as a fuppofed miracle, and not a natural caufe.

y   Nor, as others have juftly faid, can any ufe be conceived in nature for fuch a being as matter, thus defcribed. It could be only an incumbrance in creation; without thought, fenfe, motion, or agency, it could exift to no end but to employ the active and intelligent powers of nature in the humiliating office of moving to and fro, inceffantly, an uncouth, unyielding, unconfcious, and unanimated produ&ion, to which they could have no relation, and which all intelligent beings muft regard as the opprobrium of exiftence.

z   But when matter, and reft, are regarded as compound and relative beings, arifing from immateriality and motion, into which they may, by decompofition, again be refolved; we no longer find, in the order of beings, that monftrous incongruity—that infinite void between natures conjoined—thofe oppofite qualities in union—that univerfal death where only life is to be found—

thofe

thofe infuperable difficulties in the fimpleft facts—thofe miracles in the ordinary courfe of nature, which the contrary hypothefis creates; but we difcover confcious and unconfcious, material and immaterial, to be gradations in being, links of the fame chain which embraces and unites the univerfe.

CHAP. III.

Origin of the Orbicular Motions, which the ACTIVE *or* ELEMENTARY SUBSTANCE *affumes, as the conftructive Principle of Nature; and the Laws of the Elementary Subftance.*

THE elements of creation, by exifting in a motion, exift in an active ftate, and are fitted for performing a feries of fubordinate agencies, and carrying on the œconomy of nature.

B   But something besides motion was yet necessary to produce the forms of beings, such as they are. Motion, in the original element, was a physical power, an agency *capable* of producing new forms of existence; it was the active principle of creative life, the embryo of the future world. But it was *power* without direction, agency without order, activity to no end. Whatever existences its unguided agitations produced, movements, equally capricious, would destroy: connection, order, and permanency, there could be none.

C   To motion, in the elementary substance, it was necessary a LAW should be superadded; that its agency should be guided to some regular purpose, and its motions conspire to the production of uniform effects.

D   Without such a law, the elementary substance, left to movements altogether unguided and inconstant, would make the chaotic state, BEING " without form," that is, without any permanent form, connection, or order.

E   And the imposition of such a law, by a GOVERNING INTELLIGENCE, would be a CREATIVE ACT; it would *give being* to a new state of things, a NEW WORLD of order and beauty.

We

We call that a law which determines the *manner* of exifting proper to any being.

That modification, therefore, which we have already fhewn, obtains, in the active principle of nature, *motion about a centre*, may be called a *law of nature*, or a law of the ACTIVE SUBSTANCE; and it is a *general law*, becaufe it is that by which all material exiftence is fuftained.

But it is probable this is not a primary and arbitrary law, rather the refult of fome other, more fimple.

We may attempt the difcovery of fuch a law in two different ways; in one, it will ferve only to afford a more general principle than has been hitherto obtained; in the other, it will, alfo, afford a *new proof* of the truth of what has been already propounded.

If we affume the foregoing principles as true, and by inference from them, trace, upwards, the higher principles, in the fame analytic method by which thofe were inferred from more obvious truths; then the proof of the higher principles, we are now feeking, will reft on the truth of thofe prior ones affumed; and confequently the foregoing principles can receive, in this way, no additional

additional evidence, from the new ones, which are founded upon them only.

L   But if we can, independent of any thing already said, prove thefe higher principles by an appeal to new and diftinct evidences, and then, from thefe higher principles, in the method of fynthefis, demonftrate the LAW already deduced by the analytic method in the foregoing fheets, this will be a NEW proof of that law of revolving motions (G 235), and of all the preceding feries of ratiocination, from which it was deduced.

M   If both of thefe methods can be practifed with equal certainty, fuch concurring proofs, arifing from diftinct and independent fources, will ferve mutually to ftrengthen each other.

N   If we affume it as a fact, that revolving motions are a general law of nature, and we would infer from hence a law yet more general, in which *this*, of revolving motions, is imply'd; fuch law is to be fought as a RATIONAL CAUSE, and is by the firft rule (A 102), to be that *which contains in it the nature and quantity of the effect.*

O   The effect inveftigated is a revolving motion about a centre; and the queftion is, what law, or manner of being, in the ACTIVE SUBSTANCE,

is

is implied in the motion of that substance, about a centre.

In answer to this we may observe, that whatever motions revolve into themselves, or, circulating, return again to the same point, imply a *principle of union*, or a determination to unite.

In this, circulating motions differ from those which are merely progressive. The stream of a river, considered as issuing from its source, and terminating in the sea, gives us no idea of union. The waters continually change; it is not the same river to-day that it was yesterday; a part of the former has quitted its channel, its place is supplied by a fresh current: the mind sees no permanent object, can embrace no durable whole; the river is always one river, but in no two moments is it the same; in any given number of distinguishable periods, so many different rivers exist.

This is not the case where motions circulate into themselves; here is no change in the entire whole; no parts secede, none are added. It is always the same, always one, in each, and in every successive instant. Here we see a principle of union. It may be exemplified in the circulations of animals—of vegetables. In political bodies,
                                                a family,

a family, a state there is a circulation of agencies and of benefits, the head gives vigour to the members, and in its turn is sustained by them.

s   Thus we find that *a principle of union* is implied in the revolving or circulating movements of the ACTIVE SUBSTANCE.

T   But we may also assume *a priori* that a *principle of union* is a *general law* of nature, because we see, in fact, all the component parts of the universe are united systems, which successively combine into larger unions, and ultimately form *one* whole.

U   From this principle, so unquestionable, we propose to demonstrate the subordinate principle of revolving motions, which has, already, been inferred from phenomena.

X   The ACTIVE ELEMENTARY SUBSTANCE exists in a state of motion (N 106). If to this *motion* a principle or law of *union* be added, let us see what will follow.

Y   We may regard the principle of union as operating in the smallest portions of the elementary substance, forming atoms, or in any larger component parts, or in the whole universe.

And

And we muſt diſtinguiſh two ſpecies of union, z
a more and a leſs perfect: the more perfect is
that whereby units or wholes are formed, as
atoms; the leſs perfect is that whereby a partial
union connects theſe wholes, as in coheſion;
and we here conſider union in its moſt perfect
ſtate.

Let any portion of the elementary ſubſtance A
be given to form the moſt perfect union; and
let a principle of union be ſuppoſed to begin to
exiſt in ſuch portion, whoſe parts were, before,
uncohering and independent of each other.

The effect of ſuch a beginning principle would B
be a determination of all the parts of the given
portion, already in motion, to approach mutually towards each other; and there muſt be,
ſomewhere within the portion, a *centre of approach*, towards which all the parts, in common,
tend; which centre will be at reſt, and void of
any tendency, in any direction.

But this determination of all the parts of the C
portion towards a *common centre of approach*, tends
to the deſtruction both of the motion, and the
exiſtence of the ſubſtance; for ſhould all the parts
continually approximate, from a circumference,
towards a centre, they muſt continually penetrate

each

each other, till the portion ceased to have any extension, motion, or existence.

D   Now to preserve existence, and, consequently, motion, is the first law of the ACTIVE SUBSTANCE, as of all being. Union is a subsequent law. But the prior cannot give place to the subsequent; that which is essential, to that which is superadded.

E   Therefore all the parts of the portion of ACTIVE SUBSTANCE cannot continually obey their determination, induced by the law of union, towards each other, and towards a common centre of approach, because such an effect would be the destruction of the portion, and is therefore repugnant to a higher law than that of union, which disposes to it.

F   When the *direct* tendency of any inferior law is obviated by a higher law, the inferior law will operate *indirectly*, in the manner the nearest to its direct tendency that the superior law will permit. If a body in motion be obliquely obstructed, it will move on in a direction oblique to its first motion, and not opposed to it. The laws of society, when opposed to the higher law of self-preservation, are so far invalidated as that higher law

law requires; but the law of fociety muft be as nearly obferved, as felf-prefervation permits.

The ACTIVE SUBSTANCE cannot approach towards any point within itfelf, by reafon of the law of felf-prefervation; but it will have a motion as nearly approaching to a concentrating motion, as confifts with the prefervation of its exiftence.

In the parts of any extended fubftance, a motion, which is confidered as relative to a centre within the extenfion, muft be a motion approaching towards that centre, receding from it, or revolving about it. Now the motions of the parts, cannot, as has been fhown, concentrate; and they cannot retire from the centre, becaufe the parts have a determination to concentrate, and the law of felf-prefervation, which prevents their concentration, does not require that they fhould recede; therefore, although they do not approach, they will not recede from the centre; confequently, their motion can be no other than a revolving motion *about* the common centre of approach, *towards* which all the parts have a determination.

We faid the parts of the portion of ACTIVE SUBSTANCE could not approach the centre, although

though difpofed to do fo, by reafon of a higher law, which regards the exiftence of the portion, forbidding fuch approach.

K. But when, in confequence of this higher law (D 240), which compels the parts that tend towards a common centre to deviate from that tendency, and follow each other, in a revolving motion *about* the centre this modification has actually taken place; it gives birth to a new tendency, that now fuperfedes the operation of that firft law of exiftence.

L. For we have feen that the direct influence of the *law of union* is to infpire a tendency deftructive of exiftence (c 239); and while this tendency operates, and exiftence is threatened by a pofterior and fecondary law, the neceffity whereby things do *exift*, muft prevail, to obviate this *deftructive tendency*; and muft be conceived, in the firft inftance, actually to operate to this end.

M. But when, from this operation, the *union tendencies* are converted into a revolving motion, a *new tendency* is produced, oppofed to the *union tendencies*; and this *new tendency*, itfelf, fufficiently obviates the *union tendencies*, prevents the parts from concentrating, and fecures the portion from any danger of annihilation.

For

For it would feem, if not to endanger, at leaſt to impeach the dignity of exiſtence, if it were expofed to a continual attack, and if a fecondary law of nature were eternally at war with a primary. It is, therefore, fo ordered, that two fecondary laws fhall mutually counteract and balance againſt each other; that the danger which might arife from either, fhall be obviated by the other, and the ſtate of things preferved, without the continued exertion of the primary law of being, by a perpetual effort, to preferve exiſtence.

It has been fhown that motion is the original and natural ſtate of the ACTIVE SUBSTANCE, and that this motion required to be governed by fome law, in order to give being to an orderly and connected ſtate of things (c 234). Now there are motions fimple and motions complex, but the more fimple is, in all things, firſt in order, and out of the more fimple the more complex arifes, in order, poſterior. The moſt fimple motion is rectilineal; therefore a rectilineal motion is to be confidered as that which is the original and natural ſtate of things, and confequently that *to which all things tend*.

It will follow, from hence, that when any portion of ACTIVE SUBSTANCE in which the *law of union* operates, has, in the manner above explained (A 239, &c.), been compelled to assume a revolving motion, that is, a motion in some curve; a tendency to a rectilineal motion will continually exist, in every part of the revolving portion, and in every point of the curve which it describes, during its revolution.

Q And this rectilineal tendency will be a tendency to recede from the centre in every point of the revolving orbit, and to proceed in a tangent to the orbit, at each point.

R This tendency to recede from the centre, arising from the rectilineal nature of motion, will be a tendency continually opposed to that other, which arises from the *law of union*, whereby the parts are disposed to seek the centre.

S And these tendencies, to seek the centre, and to recede from it, being *contraries*, will, if they be also equal, destroy each other (E 103).

T And these will, in all cases, at some time, arrive at an equality; for the tendency towards the centre, called the centripetal tendency, that is, the *law of union*, operating first, if we suppose the motion approaches the centre, the tendency to recede from it, called the centrifugal tendency, will

will have its proportion to the centripetal continually increafed, as the orbit of revolution grows lefs, fo as ultimately to equal the centripetal tendency, and reftrain the motion from its central courfe, and at this point it will no longer feek the centre, but revolve about it.

Therefore, the revolution will be continued, u without any tendency to the deftruction of the portion, either by the annihilation of all parts in the centre, or by their diffipation and wafte in the boundlefs extent; it will be preferved by the EQUILIBRIUM of its own tendencies and laws.

It is in this part of philofophy that mathematics x come in with a timely and neceffary aid to demonftrate when the quantities of the two contrary tendencies are equal, and in what manner that equality arifes out of the relation between curves and their tangents; and in all other cafes where quantities and relations of quantities are required it is the province of mathematics to fupply the information fought.

But this is not defigned as a mathematical, but y a phyfical work; and to eftablifh juft phyfical principles is neceffary, before any utility can be drawn from mathematical fkill. Men may enter deeply into abftract fpeculations, and reafon with
profound

profound genius from data affumed to the moft fublime efforts of the human mind; but if no phyfical exiftences correfpond with thofe data, no advantages can arife to the general ftate of knowledge from exercifes of this kind, and they muft be regarded merely as *refined amufements of the underftanding*. To afcertain, therefore, facts, and true phyfical principles, is our firft concern.

z   And we have proved what was propofed; that if to a fubftance, in its nature motive, but in its parts without alliance or connection, in its motions without guidance or order, a *law of union* be added, fuch fubftance will begin to affume permanently revolving directions about quiefcent centres.

A   But the original ftate of the univerfe was that of an uncohering element; for, without doubt, parts muft have exifted prior to their cohefion; and have been extended, before extenfions were united; and it is certain, in fact, that the univerfe does exift fubject to a law of union; and that law muft have been impofed anterior to the actual ftate of union, that is, during the uncohering ftate of the elements; therefore, under the conditions neceffary to produce the revolving motions (z 246); confequently, fuch motions

muft

must be concluded actually to exist as the principle of union throughout nature.

And the proof is corroborated by the concurrence of two different methods in the same point. It hath first been demonstrated by necessary inferences, carried upwards, from obvious *facts*, to their rational causes; afterwards it hath been deduced from a principle, higher, but more evident and simple, by consequences traced downward, from the cause to its effects.

CHAP. IV.

The Mechanism of Cohesion, and the connecting Corpuscles.

WHAT has been said in the preceding chapter serves to afford some further insight to the nature of cohesion, in addition to what has been already said (c 175).

We have there shown that atoms cohere immediately by a junction of their substances, and not

not by any myſtical or obſcure power, nor by any *intermedia* of a different nature, nor by hooks, linked together; and we have now a principle which may ſhow us the manner in which the junction of the ſubſtances of atoms may be effected.

C   The principle collected from the reaſoning above, is this:

THAT THE OPERATION OF A LAW OF UNION, IMPOSED UPON A MOTIVE, BUT UNCOHERING SUBSTANCE, IS TO PRODUCE A QUIESCENT CENTRE OF MOTION WITHIN THE PORTION WHERE SUCH LAW PREVAILS, AND REVOLVING OR CIRCULATING MOTIONS OF THE SUBSTANCE ABOUT THAT QUIESCENT CENTRE.

D   And theſe motions about a quieſcent centre conſtitute the revolving ſubſtance an united and refiſting maſs.

E   The *law of union* muſt have been ſuch as not only to conſtitute the univerſe *one* ſyſtem, but to give diſtinct unities to innumerable component parts. There is the moſt ſimple and primary union of atoms, there are intermediate unions of ſyſtems, of various orders, and there is an *ultimate* union of the whole. There are unions within unions, wheels within wheels.

When

When two atoms come in contact, or the boundaries of the revolving motions of each touch in any points, a difpofition to union may arife; that is, the law of union may operate between the fubftances of each atom.

The operation of the law of union being, in all cafes, to caufe the parts in which it operates to revolve about a quiefcent centre of motion (c 248), it follows, that if parts, refpectively, of two atoms, by contiguity with each other, or by any other relations, become difpofed mutually to unite, a new centre of motion will be produced between the contiguous parts of the two corpufcles, and the contiguous parts of each atom will revolve about this new, intermediate, and common centre.

This new centre, and new revolution about it, will be a fmaller and intermediate mafs or corpufcle, compofed of parts of the two corpufcles which it ferves to unite, and it may, from its origin and office, be called a *connecting corpufcle*.

In the formation of fuch a connecting corpufcle between two primary corpufcles, when the revolving fubftance of each primary arrives at the part where the new centre, of the connecting corpufcle is formed, it deviates a little from its

K k regular

regular courſe, into the new orbit, revolving, at once, about its own proper centre, and the new centre formed between the two atoms.

к Let A. B. fig. 3, be two atoms approaching in or near contact with each other; and at the points of approach let the ſubſtance of each atom have, beſides its diſpoſition to union with the other parts of its own atom, a diſpoſition, alſo, to unite with the ſubſtance of its contiguous atom.

l Since parts, diſpoſed to union, tend to a centre, and revolve about that centre, forming, by that revolution, an united and reſiſting maſs (d 248), therefore the little maſs (M. fig. 3,) will be formed between the two atoms, out of parts of the ſubſtance of each atom conjoined.

m The maſs M. fig. 3, will, therefore, belong and be united to each atom A. and B. It will be itſelf an united maſs (u 182), laying hold of, and attached to A. and B. (m 177, a 183) and uniting them to each other.

n And although A. and B. are not united immediately to each other (y 182), but through the medium of M. yet M. is itſelf an atom, and is
united

united immediately, by interfection and combination of parts, to A. and B.

Although M. is called an atom, becaufe, like other atoms, it is formed by its own revolving motion, it is not a primary atom or corpufcle, like A. and B. but ftands in a peculiar relation to them.

M. has no feparate and independent exiftence, but fuppofes the exiftence of two prior atoms, each of which contributes its part to the formation of M.

M. alfo, is not formed by parts which revolve folely and originally about its own centre, but by parts of A. and B. which, originally and primarily revolve about their own refpective centres, and which, in the courfe of thofe revolutions, by a deviation or aberration, in a certain part, defcribe alfo the orbit of M. while they ftill continue their own revolutions.

The fubftance, therefore, of M. is continually changing, as all the parts of A. and B. near their circumference, pafs in fucceffion through M. Thus let d. e. f. fig. 4. be portions of A. revolving towards M. and g. h. i. portions of B. alfo revolving towards M.—thefe parts d. e. f.

and

and g. h. i. of A. and B. will, in fucceffion, arrive at M. and, revolving about its centre, become parts of M. till, in purfuing their courfe, they leave it, and new parts fucceed.

S   M. therefore, being formed of continually fucceffive portions of A. and B. has no proper being of its own, and may be fitly termed a CONNECTING CORPUSCLE or atom.

T   An imperfect illuftration of thefe may be had by conceiving two veffels of water to be connected by an horizontal tube, through which the two bodies of water join. The water in the horizontal tube may reprefent the relation and office of the connecting corpufcle, as it is formed out of the two bodies of water, and ferves to connect them.

U   In fpeaking, before, of cohefion, we fhowed that it muft neceffarily be produced by an intermixture of parts of the primary and moft fimple folids, becaufe the medium or uniting efficacy muft have an union in itfelf, and the parts of this intermedium muft lay hold of the parts of the atoms. But atoms are the moft fimple unions, and we ought not to devife a more complex union to connect the more fimple; therefore there is no foreign intermedium between atoms,

atoms, but they are united directly to each other; and since whatever unites two atoms, lays hold of the parts of each, two atoms, in cohering, do mutually interfect and lay hold on each other.

We showed that this could not happen with the impenetrable atoms of the moderns, nor could any rational mode of cohesion be devised for them; but that it was consistent with the nature of atoms, considered as resisting only, but not impenetrable, that such should actually have portions intermingled.

And herein we avoided all mystery, and all names of ambiguous import, and, equally, all hypothesis and invention, adhering to the most simple facts, the most general analogy of appearances in nature, and the most easy conception of the mind, that if parts of two solids combine in any small portions of their substances, the solids will adhere.

But still, a difficulty remained to be surmounted; and this was, how the atoms which were primarily formed separate and distinct solids, should be brought, afterwards, to join together; for, it being admitted that it was the nature of an atom, although not to be impenetrable, to

resist

refift penetration, fome power, or agency became requifite in order that any fmall contiguous portions of atoms fhould penetrate each other, in the manner neceffary to cohefion; and to have been compelled to feek a new caufe for this fole effect, would have been inconfiftent with the fimplicity of nature; nor do we know, in nature, where fuch a caufe could be found.

A But as, in the firft part of this work, it was faid, that the cohefion, between two atoms, ought to be confidered as an effect fimilar to the cohefion between any two parts of the fame atom; accordingly, it has now appeared that the very principle by which atoms are formed, ferves alfo, equally, for their fubfequent cohefion.

B And the difficulty above mentioned is removed, when a power already admitted to form the corpufcles, and the law in which that power operates, ferve alfo for their union, without introducing or fuppofing any new agency or new law whatever.

C The law of union, which, in diftinct portions, produces the orbicular movements that become corpufcles (c 248), operates, alfo, at the circumferences

ferences of the corpuscles, between a part of one corpuscle and a part of another.

And, as, in any portion, forming a single atom, D this law first tends to direct all the parts towards a common centre, and then to circumscribe that centre (B 239, H 241), so also, in two atoms, the parts are, by the same law, directed first to a centre between them, and then circumscribe that mean centre, which, thus, becomes the centre of union.

The same physical power, the motion of the E ACTIVE SUBSTANCE, and the same law of union given to that motion, first serves to *form*, then to *connect* the atoms.

And herein I have, as I think, followed only F the necessary deductions from self-evident *data*; and with every part of this reasoning, from the premises to the conclusion, facts and sensible evidences appear to me to concur.

From the mode of cohesion here laid down, G *varieties*, almost without end, may be distinguished, abundantly adequate to the varieties found in nature in the genera and species of things.

And

H And from the perpetual motion and activity, fhown to exift in the interior of bodies, the chemical proceffes of nature and art, vegetation and animal life, find their fource. The changes which bodies undergo, fpontaneoufly, or by admixture, gradually or fuddenly; the commotions, agitations, and fucceffive forms of nature, are found to originate in the conftitution and conftruction of matter: but we leave thefe things to a future time, and challenge the affiftance of others in cultivating fo vaft a country and fo rich a foil.

I One hint only I fhall throw out, for the chemifts. Since every fingle atom is a little revolving fphere, if two atoms only are joined to each other, by a connecting corpufcle, both muft revolve about their common centre, as the moon and earth are carried about their common centre by the revolving active fpheres of each; and the fmaller of two corpufcles will be carried about the larger, or feveral fmaller corpufcles may be carried about one larger: and fince, in fluids, the cohefion is lefs than in fixed bodies, and the corpufcles are held together by lefs firm or lefs numerous attachments, this ftate, of fluidity, is favourable to the various changes of decompofition and union which happen in fluids with fo much rapidity; and the inteftine commotions

motions which appear during chemical proceſſes, may, moſt of them, ariſe from the looſe corpuſcles carrying, or being carried rapidly about the centres of others, according to their ſeveral relative dimenſions, denſities and other circumſtances; while fixed bodies, having their corpuſcles, on all ſides, mutually and more firmly connected, allow leſs of theſe inteſtine motions, each corpuſcle having ſo many and firm attachments, as to prevent it from being carried about the orbit of any one, or from changing its ſituation in the maſs to which it belongs: hence, chemical changes happen more ſlowly in fixed or ſolid bodies.

CHAP. V.

Of the Orbicular Motion of the ACTIVE SUBSTANCE *in Mechanical Impulſe.*

WHEN we ſpeak of the viſible motions of our own, or neighbouring bodies, we conſider the earth as being at reſt, and regard only

only the local, relative, and tranfitory motions, with their local caufes.

B Thefe tranfitory motions, of which furrounding bodies are the fubjects, we afcribed to caufes equally tranfitory; to an adventitious principle, of a motive nature, whofe occafional prefence moves the bodies, and whofe exit leaves them quiefcent, as before.

C Thus we found a fubftance which freely pervades matter, and which, efcaping from every body which impels another, enters the body impelled; and, in fucceffive impulfes, fucceffively tranfmigrating, as it continues to flow, till it ultimately arrives at its common refervoir, the earth.

D In every impulfe, whether preffure or percuffion, the occafional caufe flows through the bodies concerned in the effect (B 130).

E Hitherto we have fpoken only of impulfe in its moft fimple form, or of a fingle impulfe, wherein one body is the agent alone, the other only the patient; one the tranfmitter, the other the recipient; and wherein the flowing fubftance paffing through and out of a body, returns to it no more.

The

The fame principle may now be applied to ᴦ cafes of compound impulfe. If a body which fuftains another's preffure, be alfo, itfelf, a preffing body on that other, the principle of the flow of the ACTIVE SUBSTANCE becomes applicable to both; each is reciprocally agent and patient; each communicator and communicant.

While the one, as an agent, is parting with ɢ its ACTIVE SUBSTANCE, it is, at the fame time, as a patient, receiving that which iffues from the other.

The mutual preffures being fuppofed *direct*, ʜ each body will now become the fubject of two contrary currents of ACTIVE SUBSTANCE, each current impelling the body in the direction of its own flow.

Confequently, each body will, in mutual and ɪ direct preffures, be at the fame time impelled, or rendered active, in two contrary directions.

If the preffures, therefore the currents, there- ᴋ fore the activities, are equal in contrary directions, the body will be active ultimately in no direction (ᴇ 103), and will remain at reft.

L l 2   Agreeably

L   Agreeably to which the fact is, that bodies which prefs, and fuftain preffure, equally, in contrary directions, are not moved thereby.

M   Every current muft have a fource and a termination.

N   In mutual and contrary preffures the contrary currents have, reciprocally, the fame fource, and the fame termination. For wherever the preffure in one direction is conceived to begin, there it muft be conceived, the current in that direction originates. And as in every part of one current, the contrary current counteracts and deftroys its influence, the contrary current muft proceed from its own fource, to the fource of the other, and there have its termination; and this reciprocally.

O   If two contrary currents have, reciprocally, the fame fource and termination, that is, if the fource of one be the termination of the other, reciprocally, the two currents, with their fources and terminations, form together one circulatory fyftem, revolving into itfelf, and thereby conftituting the feveral parts one entire whole.

P   Therefore, in all contrary impulfes, however tranfitory,

tranfitory, a circulatory motion of the ACTIVE SUBSTANCE, revolving into itfelf, is produced.

Such a motion has been fhown to be the con- Q ftructive principle of matter and of the world. If, therefore, the fame principle by which the permanent appearances of nature are produced, obtains and operates in the accidental occurrences of contrary preffures, there ought to be found a perfect analogy between thofe permanent and thefe accidental effects, which depend thus on fimilar caufes.

And this analogy is actually found. R

For the operation of the orbicular motion in S atoms, or in the larger fpheres (D 248), is to produce an UNION among parts, otherwife diftinct. An atom confifts of parts, which, independent of their motion, are uncohering, and divifible, without end; but, by their orbicular motion, they together conftitute one undivided atom.

In like manner, two bodies, reciprocally pref- T fing on each other, and thereby, as has been fhown, *being contained within the courfe of one circulatory fyftem*, or revolving motion (O 260), become

come abfolutely one mafs, fo long as the preffure continues.

υ For, to have parts held together is the characteriſtic of unity. If *two* femi globes be attached together at their bafes, they become *one* globe. It is by various forts of attachment, that *feveral* parts are made to form *one* object, a building, a machine.

x Mutual preffure attaches bodies together; while the preffure fubfifts, they cohere, are held to each other, they are conjointly moved or at reſt, they are infeparable, unlefs a force is employed greater than that which holds them together, and which, in all cafes, diffolves unions. The union is, therefore, perfect between two bodies which mutually prefs and are preffed.

y The analogy is therefore perfect between any fuch tranfitory preffure, and the more permanent effects of nature.

z As from the fimilarity of the principles, operating in thefe different cafes, we *inferred* the analogy which does in *fact* appear; fo from the analogy, in itfelf obvious, we might, with equal certainty, have inferred, by a contrary method, the fimilarity of the operative principles.

The

The scene now expands upon our view; we have traversed from the primordial parts, to the limits of the world, and found nature, amidst all her infinite diversities, still uniform. We are now able to add to this catalogue innumerable, intermediate, local, and transitory appearances; among which are all the effects of art, wherein action is mutually sustained and exerted: in all of these the active principle performs that circulatory flow by which every union is produced and sustained.

CHAP. VI.

Absurdity implied in the Question concerning the Origin of Motion.

THE question, whence is the origin of motion, supposes that rest was the primitive state of matter, and that motion was produced by a subsequent act. But this supposition must ever be rejected, as it is giving precedency to the inferior, and inverting the order of nature.

What

B   What life is to death, motion is to reſt. Was death, then, the firſt act of creation, and did life ariſe from that death? Was death the immediate offspring of Deity, and life produced in a ſecond generation? Or had death exiſtence from eternity, and, in time, did life iſſue from its womb?

C   If the firſt creation were reſt, the ſubſequent act, that produced motion, deſtroyed the reſt, which was firſt born; or if reſt were eternal, that eternal ceaſed to be, when motion came into exiſtence. But wherefore did the Creator produce, what by a ſecond production he muſt deſtroy? or how could non-entity annihilate its eternal rival, and ſtart into its vacant throne of being?

D   But to ſuppoſe reſt prior, and motion ſubſequent, involves all theſe incongruities, and the queſtion which implies this ſuppoſition, is, therefore, founded on a fundamental error, and ought, in the firſt inſtance, to be rejected.

E   Our ſenſes ſtill deceive us, and notwithſtanding the high rank experimental ſcience now holds, and the infallibility aſcribed by ſome to the teſtimony of our *nerves*, it has ever been the office of reaſon to correct the errors into which an implicit confidence upon their report precipitates us;

us; and this, our intelligent guide, finds her labours multiplied before her by the well-meant endeavours of many, in the prefent day, to difpenfe with her aid.

The apparent reft of bodies around us we miftake for real, and then inquire whence the motion is derived that we perceive to fucceed that reft. But all thofe bodies, which appear at reft, are already in motion; and thofe motions which to us feem begun, and fucceeding to reft, are no other than certain tranfitory inequalities, which arife from local caufes in the diftribution of the general current of motion, by which all bodies are inceffantly carried along, excepting fuch irregularities, with an uniform flow.

We miftake the world for a ftagnant lake, in which, while we appear to float at reft, we inquire from what caufe neighbouring bodies occafionally move.

But we are, altogether, floating along a current which is imperceptible, becaufe we are carried with it, and no banks enable us to diftinguifh its flow: while all things in view move uniformly along all feem at reft; but if, here and there, one of the floating bodies is hurried on before, or detained behind its companions, this inequality gives

gives the appearance of a motion begun, and we demand its fource, fince all around is reft. But if we knew or recollected that, in fact, all is motion around us, we then fhould not inquire, in any one body, the fource of a motion which it has in common with others, but the fource, only, of fome increafe or decreafe of motion which diftinguifhes it from its neighbours. When a body, as we fay, is moved, it is not *then moved*, but only fuffers a change of the motion it before had, by communication with local caufes which render the general diftribution unequal.

It was the apparent reft of bodies, and the appearances of beginning motions, that firft led men to feek their origin, as thinking it an accidental mode of being, fubfequent to the exiftence of matter: and though the ground of this inquiry was deftroyed, when the motion of the earth was difcovered, yet the fpirit remained; for ftill the earth was fuppofed to have exifted once at reft, and the queftion of the origin of motion was transferred to it, placing the inquiry on a worfe footing than before. Men were, perhaps, cautious of bringing in the direct agency of the Deity, to move fmaller bodies, but when the earth and planets were in queftion, they thought the bulks and the velocities of thefe bodies were the effects of a power not unworthy

worthy an Almighty arm; and principles, admitted, depriving them of any other caufe, philofophers fcrupled not to introduce the Creator in the world already formed, launching forward, by a mighty impulfe, the planetary orbs, in their feveral fpheres. A magnificent thought for the heathen mythology; worthy the arm that fubdued the Titans, or that which, by one ftroke, could open the ponderous earth to its very centre ; but infinitely unworthy that INTELLIGENCE from which the univerfe arofe.

The incongruities which attend the queftion, how motion began, are not retained, if motion be conceived as original, and reft as its produ&ion ; for although, if reft were original, motion would imply its deftruction, yet if motion be primary, the reft which fucceeds, does not imply the deftru&ion of the motion, it being only certain relations of the motion agreeing with it and produced by it (o 230). Nothing, then, once created, is deftroyed; and motion, as it is more excellent, is precedent to reft. Further, if reft be the firft quality of things, it muft be an abfolute reft—an entire reft—a negation of motion—an effential death : but if reft be only a production of motion, it is then only a relative reft, reſt, ſuited to the nature of a living world, and ſuch a reſt as alone does, in faƈt, exiſt.

CHAP. VII.

Law of Union, in what Reſpeƈts agreeing with the Attraƈtion of the Newtonian Philoſophy, and wherein diſtinguiſhed from it.

A THE law of union, above ſpoken of, has in ſo much, an affinity with the modern attraction, as to render a comparative view of theſe two principles expedient.

B Attraƈtion is a name given to a tendency to approach each other with which all bodies are ſaid to be endowed, in the direƈt proportion of their maſſes, and inverſely in the ratio of the ſquares of their diſtances.

C The law of union, of theſe pages, ſaid to exiſt, not in matter, but in the elementary principle of matter,

matter, is a tendency, in certain parts, and portions, to approach each other; but in what ratio to the magnitude and diftance, has not yet been faid.

Thefe two principles agree, in that both affert a principle of union to exift in nature, and a difpofition to approach, to be given to diftant beings, by a governing power.

They differ in that one fuppofes this principle to be impofed immediately upon matter; the other fhows the principle of union to be given to an immaterial and primary element. Let us inquire to what this difference leads.

The principle of attraction is employed to explain the motions of the celeftial bodies. In order to this, it fuppofes, befides the attraction, two things, firft, an impulfe given to thefe bodies; fecondly, an innate principle in the bodies continuing the tendency, the impulfe imparted, towards a rectilineal motion. The attraction, as a third force, is employed to draw the bodies from their rectilineal direction towards a centre, and from the balance between the centrifugal and centripetal tendencies, their revolutions are found to refult.

Although

G Although this result be granted to flow from these things supposed, yet the suppositions are inadmissible, partly as being hypothetical, partly as being inconsistent with themselves, or with other principles received. An impulse miraculously given to the planets after they were created, is merely hypothetical; an innate principle of motion in the bodies, whereby they continue to move after the impulse, is inconsistent with itself; for the necessity of an impulse to move the bodies, supposes them not to have any innate principle of motion. A tendency of bodies to approach each other, is inconsistent with another principle received, namely, that bodies are inactive; for, if inactive, they can have no tendencies, nor can act upon each other.

H The law of union serves to a much more extensive purpose than the attraction of bodies, while it is neither hypothetical, nor incongruous. Attraction presupposes the existence of matter, the law of union explains its formation.

I The law of union is a tendency said to be imparted to a substance already active, and endowed with tendencies; it therefore implies no contradiction.

It

It enables us to explain the motion of bodies K
without fuppofing a preternatural and miraculous
impulfe given to the planets, an event merely
imaginary, folitary, unparalleled in nature, and
inadmiffible; and alfo without requiring that
other principle in bodies, in itfelf fo contradictory and incomprehenfible, the *vis inertiæ*,
faid to be indifferent to motion or reft.

From the law of union alone, exifting in the L.
ACTIVE SUBSTANCE, all thofe confequences refult, to which attraction is infufficient without
two other concurring hypothefes: it ferves alfo
to much more extenfive purpofes; it is more intelligible, more fimple, and more ufeful.

If it be faid that, neverthelefs, the law of union M
is the fame thing with an attraction, if applied to
the parts of the ACTIVE SUBSTANCE, I reply,
that I think the term attraction, however it may
be defined, exceptionable from its etymology and
its general acceptation, becaufe it implies not
merely a tendency between two diftant beings to
approach, but rather an agency between them,
mutually exerted on each other, by pulling or
drawing. But fince, when I fay that parts of
the ACTIVE SUBSTANCE actually tend to unite,
I do not mean to affert that one part draws

or

or pulls another; I think it important to lay aside the term attraction, as being liable to misinterpretation, and having been, in fact, a source of much error.

I have hitherto termed the active element of the world the ACTIVE SUBSTANCE, by this denomination to distinguish it from matter, which is an inactive substance; and have avoided even terming it a fluid, lest this term might convey some idea of materiality. It is, however, the original and most perfect fluid, without being material, since it possesses essentially, that requisite, a facility of motion among its own parts, which gives an imperfect fluidity to matter.

PART

PART IV.

Some Abridgements of the foregoing Ideas; their Agreement with Facts: Conclusion.

CHAP. I.

Constitution of the Universe.

THE original state of the universe is that of a most subtle and perfect fluid (N 272).
This fluid is not formed of parts, each, in itself, a solid, and becoming fluid, in an aggregate mass, by their want of attachments, and by their smooth surfaces admitting of a free motion among each other; but it consists of homogeneous parts, every

where

where unfolid, and devoid of attachment; the fmalleft portions being fluid equally with the larger, and with equal facility admitting of motions among each other, or each within itfelf.

B Motion is an effential property of this primitive fluid.

C The motion of the primitive fluid is a force, agency, or phyfical power.

For by force, agency, or power, is fignified whatever can produce a change, or a tendency to change; now a fluid, to which motion or change is effential, can, by union with other beings, induce on them its own changes or difpofitions thereto.

D This moft perfect fluid, whofe parts are no where held together, cemented, or attached quiefcently to each other, is, by its motion, formed into folid exiftence, with parts united, firft, within themfelves, and fubfequently to each other.

E The motion whereby the primitive fluid is conftituted a folid, is an orbicular motion, or a progreffion revolving into itfelf.

For

For all thofe parts, otherwife difunited, which continually revolve in the fame orbit, about the fame centre, are united in that orbit, and by that centre; they preferve the fame diftances and relations, and conftitute one whole. The parts of that whole are forcibly held together by their determination to continue in the fame orbit; and the whole, fo formed, is a folid or cohering mafs.

The folid, thus produced by the rotation of a portion of the primitive fluid about a centre within the portion, is at reft, excepting only, whatever motions it may have, induced by external caufes.

If two fuch folid fpheres interfect each other, or are each interfected by a third, they will hold together, or cohere.

The univerfe is a congeries of fuch folids, of different denfities and magnitudes, of which fome are held together by interfection, others, fmaller, are contained within larger, in feveral fucceffive gradations.

The orb of the folar fyftem is the largeft folid with which we are acquainted; each of the fixed ftars have fimilar furrounding orbs; and thefe orbs,

orbs, by mutual interfection, cohere together, forming, by this means, the quiefcent mafs of our univerfe.

K   Within the orb of the folar fyftem are contained feveral fmaller folids, to wit, the planets primary and fecondary, which are found, by obfervation, to revolve about the fun, and, together with it, about a common centre.

L   Thefe folid bodies being contained within the orb of the larger folid, the extenfion of the folar fphere of activity, muft neceffarily be carried about the centre of that larger orb, by its revolving motion, at the feveral diftances wherein they are placed.

M   Befides, the bodies of the planets, there are lefs denfe folids, which encompafs their more firm maffes to a certain diftance. The atmofpheres of the planets are larger folids, enclofing the planets, as nuclei; but ftill larger, are the active fpheres of the primitive fluid, which envelop each planet.

N   Several of the planets, befides being involved and carried round in the fphere of the fun's activity, have their own particular attachments. Smaller planets have their active fpheres attached

to

to the active spheres of larger, and, by the larger active spheres, are carried about the larger bodies, at the same time that they accompany the progress of these larger, about the sun. Such are the several secondary planets.

The active spheres of any planet, of the sun, O or of any fixed star, are, each, *single solids*, attached to others.

The body of the earth is a mass, or *congeries* P of smaller solids, attached together.

Of these, two kinds are united in different ways. Q

First, there are loose or uncohering bodies, R animals, stones, fluids, &c. moveable on the earth's surface, but held and tending to the earth, and if forced from its surface, returning to it when left to themselves. These loose bodies are, by the action of the external sphere, confined in one mass, making the globe of earth, which is a central nucleus to that external solid, or circumscribing active sphere. This tendency of all parts of the earth to its centre, is denominated weight, or gravity, and is commonly explained by a supposed mutual influence between all parts of the earth, whereby the general

ral mafs detains each particular body, or draws it back, if by any means it has been removed; in fine, by a law of union impofed on all matter. But matter being a folid, formed by fuch a motion of the primitive fluid as renders the folid inactive, is incapable of receiving any law of union, becaufe fuch a law implies activity. Matter, therefore, can only receive activity in a fecondary way, as fuperadded by the primitive and active fluid, and acquiring fuch tendency, motion, and direction, only, as this fluid imparts, itfelf being a pure, *unrefifting* paffivity.

s   *Secondly*, befides this attachment from without, by which all parts of the earth, however unconnected in themfelves, are compreffed together, by a furrounding fluid, there are immediate and internal attachments of parts to parts, which exift independent of that extraneous force.

t   Thefe are the cohefions among the fmaller and primitive folid fpheres or atoms by mutual interfections, which preferve the maffes of fenfible bodies, in the various forms which art and nature prefent to us.

u   There are two forts of primitive folids; the primary atoms, and the larger fpheres of the univerfe. Thefe refemble each other in their

laws

laws and conſtruction, differing only in magnitude and in denſity.

There are two ſorts of maſſes conſtituted by x the interſection or coheſion of theſe ſpheres, ſenſible bodies, and the maſs of the viſible univerſe formed by the coheſion of the ſtarry ſpheres.

The ſmaller maſſes, of ſenſible bodies, formed y by the attachments of the ſmaller ſpheres being contained within the larger order of primitive ſpheres, are, by the law of union of thoſe ſpheres, driven each to its own centre, and acquire a new union, to wit, of gravity, by the ſuperincumbent and ſurrounding influence of the larger primitive ſpheres. The ſmaller ſpheres, or atoms, and the maſſes they form, being all thus driven to the centre of their reſpective larger and containing ſpheres, form the globes of the ſtars, ſun, and planets, each in the centre of its own ſphere.

The ſubtle revolving fluid, the centre of whoſe z vortex the earth occupies, not only ſurrounds, but pervades the earth, and other vortices their earths, to their centres; and the earth and planets are, by its revolutions, carried around on their own axes.

The

A   The earth is an inactive mafs, and all its component maffes are feverally, as collectively, inactive; but the earth, and all its parts, have various collective and feparate movements, imparted from the fluid which furrounds, pervades, and conflitutes it.

B   Being immerfed, together with its proper furrounding fphere, in the larger fphere of the fun, it is carried thereby, in a large orbit, about the fun.

C   At the fame time, by the revolution of its proper fphere, it rotates on its own axis.

D   Thefe two motions belong to the whole globe, and by thefe, the component parts, at equal diftances from the centres of motion, have equable velocities.

E   Befides thefe, parts of the earth have various relative motions among each other, fuch as currents of air and water, earthquakes, the ordinary phenomena of animal and vegetable life; of mechanical or chemical changes ; which occur to daily obfervation.

F   The globe of earth is immerfed in two diftinct oceans of the pervading fluid; its own proper fphere, sphere, and that of the sun. Besides these, it consists of the same fluid, modified, by revolving in smaller spheres, into the primitive material atoms.

By certain relative circumstances, the active fluids, which pervade the earth, may be determined to some local irregularities, so as to impart to certain bodies, more or less than their common proportion of activity.

The primary atoms are formed inactive by the orbicular motion of the active fluid about a centre, that is, in a ma..~~r, by the folding up of the active fluid, into itself. If, by any processes, or mutual agencies, any of these lose their orbicular motion, and are unfolded, so that the fluid, instead of revolving in the small orbit of the atom, shall move progressively onwards, the atom is then *resolved* into the primitive fluid; and the portion of fluid which served to form the atom, now resumes its original active state, and becomes a motive agent to other atoms or masses of matter with which it may unite.

It is probable such an unfolding or decomposition of the primitive atoms takes place in nature, and that herein are found the origin of many natural agencies; namely, in matter, itself, restored to its priſtine active ſtate; and that from theſe, and from the cauſes above mentioned (G 281), all the apparent motions of bodies upon the earth, all changes, proceſſes, and productions, are to be derived.

K   When, from either of theſe ſources (G, H 281) any atom or maſs becomes impregnated with more than its common proportion of the active fluid, or with portions of the active fluid directed differently from their ordinary courſe; or any portions, by communication with others, loſe part of their common proportion; ſuch atom or maſs, which before had a relative reſt among its neighbours, will now have an *extraordinary*, and thence an *apparent* motion; and ſome change will thence be produced in the ſenſible appearances of bodies.

L   And in this way do the various phenomena of nature and art occur, to wit, by relative motions of parts among parts, induced by ſome local changes in the equable diſtribution of the conſtituent or pervading fluid (D 113).

M   A more particular explication of phenomena by ſhowing their correſpondence with various relative circumſtances of this fluid, will belong to a future work.

C H A P.

CHAP. II.

Agreement of Phenomena with the Theory.

IF the theory laid down be true, propositions of facts, deduced from it, will agree with appearances in nature; and this agreement will serve especially to confirm the doctrine, because the principles were not gathered, subsequently, from experiments, and fitted to them, but have been drawn from independent sources, purely intellectual.

A BODY BEING AT REST, WILL, OF ITSELF, NEVER BEGIN TO MOVE.

Because, being inactive, it contains, in itself, no cause of motion or motive tendency.

OBSERVATION I.

A body is never found to begin to move without the accession, to it, of some external cause.

DEFINITION.

Resistence is an action considered in its relation of opposition to another action.

E A BODY AT REST WILL NOT RESIST ANY ACTION OR FORCE EMPLOYED TO MOVE IT.

Becaufe, being inactive, it is incapable of refifting, which is acting.

OBSERVATION II.

A body at reft, fuftaining any motive agency, acquires a motion or motive agency, equal to what it fuftains, the external obftacles of friction and *media* being taken out of the account.

F This fact is univerfal and agreed upon. Friction and mediums diminifh or deftroy a part of the motion, which, otherwife, the body would acquire from the influence exerted on it: thefe, then, are truly refiftences, or contrary actions: but abftracting thefe, or, if there be no fuch refiftences, the whole influence exerted, is found in the body. This could not be the cafe, if the body refifted: if it refifted, fo much as it refifted, it would, like the afore-mentioned refiftences, diminifh motion, and the refiftence of the body ought to be, always, a quantity, either by itfelf, or in addition to the other refiftences, the meafure by which the lofs of motion fhould be determined. But the lofs of motion follows the proportion, only, of the other refiftences, friction, and *media*; and if there be no friction

nor

nor medium, there is no lofs of motion; confequently, the body does exert no refiftence, nor is there *any* refiftence befides thofe already mentioned. In the falling of a heavy body, the body acquires all the momentum, which, from its weight, and the time of its defcent, it ought to poffefs, excepting fo much as muft be taken away, on account of the known refiftence of the air.

Observation III.

If a heavy body be fufpended by a ftring, G and the ftring be cut, the body will defcend.

As in this example there are not the circum- H ftances of lofs of motion by communication, and the fhock attending impulfe, the fact is more fimple and conclufive; the motive influence is invifibly, filently but continually exerted, and the body, although at reft, conftantly tends to move. The feparation of the cord which fupports it, is not an action upon the body, it is only the removal of the contrary action by which it was held in equilibrio: if the quiefcent body, when its fupport was removed, *refifted* the influence which determines it to fall, this refiftance muft be in fome way manifeft, and muft impede the fall proportionally to fuch refiftence, at the very commencement of the motion; but fince no fuch impediment appears, the fact demonftrates, ſtrates, that no reſiſtence whatever is exerted by the body.

I IF THE FORCE EMPLOYED TO MOVE A BODY AT REST BE AN IMPULSE, THE IMPELLING BODY WILL LOSE AS MUCH MOTION AS THE OTHER ACQUIRES; ABSTRACTION BEING ALWAYS MADE OF THE RESISTENCES OF MEDIUMS AND OF FRICTION.

The impelling body being actuated by an adventitious fluid, which, in the impulſe it communicates to the body impelled, muſt itſelf have ſo much motion or activity leſs, as it has parted from to the other (R 127).

OBSERVATION IV.

K Let a body, at reſt, ſuſpended freely, and inelaſtic, be ſtruck by another inelaſtic body, of an equal maſs, both bodies will proceed together with half the velocity of the impelling body.

L The impelling body, after the ſhock, having only half of its former velocity, hath loſt half its motion, and the other hath gained an equal quantity; its motion from a ſtate of reſt, being, after the ſhock, equal to what the impelling body hath loſt.

The

The diminution of motion in an impelling body has been attributed to refiftence in the body at reft, but, on *any principle*, this is inconfiftent with the *fact*; for, if a body in motion, agreeable to the received theory, moved and impelled by an *innate power*, and it was *refifted* by a body at reft, the impelling body would be impeded by the refiftence, but the body at reft would not be moved for two reafons; *firft*, becaufe it could receive from the other none of its innate power, by which it had impelled; *fecondly*, becaufe that power of the impelling body, which is refifted, cannot produce its effect, at the fame time that it is refifted; but the body at reft receives power from the other, therefore the power is not innate, and the power of the impelling body has its full effect, therefore is not refifted. The diminifhed motion of the body which impels, is an effect of its having parted with its motive influence: it is fo much diminution, in confequence of fo much deprivation.

As the body at reft *cannot* refift becaufe it is *inactive*, fo the body which lofes its motion, does not *require* to be refifted, becaufe it has loft its difpofition to move, at the fame time that it has loft its motion. The fupport of this doctrine of *refiftence in a body at reft*, lays in a mifufe and equivocation of terms.

Bodies

O   Bodies are said to be inactive, and to resist by their inactivity. Resistence, therefore, is here used in a sense seemingly contrary to ours; it is not an action, but belongs to inactivity. So absurd a thing as resistence without action might shock our reason; but this absurdity is apparent, not real; the use of terms only is reversed, inactivity means activity. The inactivity of body, and the resistence of inactivity, are in truth the activity of body, and the resistence of activity; for this inactivity is said to *be exerted, to persevere, to impel, to react, to produce change in other bodies, to be a power or force*, and all these, more or less, proportionally *to the body whose force it is*. If these are not active qualities, I beg to know what are? If inactivity be said to perform and be all these, what is the difference between it and activity?

P   Such is the manner in which this term *inactivity* is perverted and abused to answer the purposes of an hypothesis.

Q   Definitions are the basis of all demonstration: the terms activity and inactivity are not defined in the received system, but they are confounded together.

This

This play, on terms, may again be obferved, in the method of proving, that a body in motion, perfeveres in motion by its inactivity. For, it is faid, that it requires an action, to ftop a body in motion; and fince a body in motion, is inactive, it cannot exert an action to ftop itfelf; it muft, therefore, go on to move for want of action to ftop itfelf, that is, by inactivity. There is an inconfiftency throughout this argument. A body in motion requires an action to ftop it, this is evident; but, is this a proof that it is inactive? On the contrary, furely, it is a proof of its activity, while in motion. An action is required to ftop it; but, it cannot be expected, that the fame activity, which moves the body, fhould alfo ftop it. The body goes on, not for want of activity, or by inactivity; but becaufe its activity difpofes it to move, and not to ftop.

A BODY AT REST, BEING IMPELLED BY A MOVING BODY, RESISTS THE BODY WHICH IMPELS; ALTHOUGH IT RESISTS NOT THE FORCE OF THE IMPULSE.

This is a very important diftinction, between refifting the *mafs* of a body, and refifting its *motive* force: and we fhall find it correfpond throughout, both with theory, and with fact.

u   All refiftence is action, and fuppofes another action, refifted. It is denied, that a quiefcent mafs refifts the motive force of another mafs impelling, becaufe the quiefcent mafs is inactive, and the force being tranflated only, is not, in fact, refifted. But, if a quiefcent mafs refift the mafs and not the motive force of an impelling body, there ought to be an action in the quiefcent mafs; and, in the motive one, an action diftinct, from that by which it is motive; fuch, in fact, are the conftituent activities (D 113), or cohefive powers of every body. Thefe actions are alike in a body, whether it be at reft, or in motion. They preferve the forms of bodies, and hold their parts together. In impulfe of a motive, on a quiefcent body, the motive body, endeavouring to continue its motion, becomes a force, whereby the cohering mafs tends to penetrate the quiefcent body, and this tendency is an action on the cohefive force; the cohefive force refifts, and by this refiftence, acts alfo on the impelling mafs to penetrate it. Thefe are actions and refiftences of the conftituent, or cohefive activities, and have no effect on the motive activities, or the motion of the maffes.

Observation V.

x   A mafs in motion, has its velocity diminifhed, by impulfe on a quiefcent body.

Refift-

Refiftence is an action, which impedes, or ob- Y
ftructs the effects of another. The retardation
of a body, is an evidence that the body has been
refifted, as the communication of motion fhows
that the action has not.

A BODY ACTED ON WHEN AT REST, DOES Z
NOT RE-ACT.

Becaufe it is inactive; nor can action proceed
from reft.

OBSERVATION VI.

If a body at reft receives an impulfe, it ac- A
quires fo much motion as the impelling body
lofes.

EXPLANATION.

If the body at reft re-acted, it would deftroy B
the force of fo much of the impulfe, as was
equal to the action, while the re-action would
alfo be deftroyed, in the manner that contrary
actions oppofed, always deftroy each other; and,
confequently, if the re-action were equal to the
action, both would be deftroyed; but as there
is no deftruction of action, merely a tranflation of it, there is no re-action. The *re-action*
may be fuppofed to be the fame with the *refiftence*
of a body at reft; yet, they are inconfiftent
with each other, becaufe the *refiftence* is faid to be

pro-

proportional to the *mass:* the *re-action* to the *action*, which different proportions of the same power, in the same body, are impossible, inconsistent with facts, and with each other, and are together, to be rejected.

C IF A BODY IN MOTION, RECEIVES AN IMPULSE, IT DOES NOT RE-ACT ON THE IMPELLING BODY; IF THE IMPULSE BE OPPOSED TO THE DIRECTION OF THE FIRST MOTION, THE IMPULSE OR ACTION IS RECIPROCAL; IF THE IMPULSE BE NOT OPPOSED TO THE DIRECTION OF THE FIRST MOTION, THERE IS NO RECIPROCAL ACTION, AND, IN EITHER CASE, THERE IS NO RE-ACTION.

A body in motion is active by reason of its motion; it can act or impel in the direction of its motion, and, therefore, a body impelling it, contrary to its direction, is, by it, reciprocally impelled. But this action on either, is the effect of the separate states of activity induced upon each body; neither action is consequent upon the other action; therefore, neither is re-action, which is a consequent action. When the impulse is not opposed to the direction of the moving body, there is no reciprocal action; because the moving body, though active, does not act against the direction of its own motion, and, in all cases, actions are proportional to the activity

tivity of their own bodies, not to the action of other bodies.

IN IMPULSE BETWEEN TWO BODIES IN MOTION, EACH UNDERGOES AN EQUAL CHANGE IN A CONTRARY DIRECTION.

Whatever motion either body imparts to the other, itself is deprived of, and retains so much less in the same direction as the other has more; the changes thus produced, are therefore equal, and they are contrary, since one body is accelerated, in consequence of the others being retarded.

IF ANY INVISIBLE FLUID SURROUNDS A BODY, PRESSING TOWARDS IT EQUALLY ON ALL SIDES, AND BY THIS EQUALITY ON ALL SIDES, PRESERVING THE BODY AT REST; AS THE FLUID X, X, X, ENVIRONING THE BODY M, FIG. 5. AND ANOTHER BODY (n) FIG. 6, APPROACHING TOWARDS M, BECOMES IMMERSED IN THIS FLUID, AND RECEIVES THE ACTION OF SUCH PART OF THE FLUID AS FALLS ON, OR ENTERS INTO IT, THE TWO BODIES M AND (n) WILL TEND TOWARDS EACH OTHER WITH EQUAL MOMENTA, AND IF AT LIBERTY, WILL, WITH EQUAL MOMENTA, APPROACH EACH OTHER.

For

For the body (n) receiving the action of so much of the fluid as falls upon it, diminishes, on that side, in the same proportion, the pressure upon M, and, therefore, M is no longer pressed equally on all sides, but most on the side (o), contrary to (n), where the pressure is undiminished. The body M, will therefore move towards (n), by the superior pressure on the side (o); at the same time, (n) being pressed towards M, by the fluid, of which it has deprived M, will move towards that body; and the pressure on (n) being precisely that which is wanting to M, and which gives the superiority to the side (o) of the body M, the momenta of M and (n) will be equal, and their velocities reciprocally on their masses.

Observation VII.

g If a piece of iron be placed within the action, which surrounds a magnet, and which, being equal on all sides, preserves the magnet at rest; the iron, being acted upon by the surrounding influence of the magnet, both bodies, if floating in water, or otherwise at liberty to move freely, will approach each other with equal motions, and, in proportion, as either mass is less, its velocity will be greater than the other.

Explanation.

h Whatever be the particular theory, whereby all the phenomena of the magnet are explained,

it

it is inconteſtible that ſome agency ſurrounds magnetic bodies, and actuates maſſes, or particles of iron, within its reach. There can be no doubt, but that the magnet is the centre, and the ſubject of this agency; and it is certain, that it influences the magnetic body equally on all ſides, otherwiſe it would, at all times, be moved, when left to itſelf, by its own magnetiſm. It is equally evident, that any given portion of the ſurrounding activity of a magnet, cannot at the ſame time, act upon two bodies; conſequently, if, at any diſtance from the body of the magnet, any portion of the ſurrounding agency, finds a new ſubject of its action, ſo much action is taken on that ſide from the magnet itſelf, the equilibrium is thereby deſtroyed, and the action on the contrary ſide, preponderates, by the quantity removed from the other. The ſurrounding agency is, in fact, ſtill equal on all ſides its own centre; but not on all ſides the body of the magnet. The centre of this agency is now changed, from the centre of the magnet to ſome point between that centre and the piece of iron; both bodies are now circumſcribed within one ſphere of activity, and though, at a diſtance, in reſpect to their viſible ſurfaces, they may be conſidered as one united body, in reſpect to their attachment (s, т 261); and the centre, common to both, is now the centre of the agency towards

wards which both bodies move, and at which they will meet.

I The propofition, may be equally well exemplified, and the fact remarked in all other cafes of an invifible influence, between diftant bodies; as in coheflon, in electricity, in gravitation, or in all thofe phenomena to which the term attraction is applied, whether to exprefs the fact, or to denominate the caufe; but which term, either as expreffing the effect or the caufe, we have thought it expedient to avoid, for reafons above given.

K IF AN INVISIBLE FLUID PROCEEDS FROM ANY CENTRAL BODY, M, FIG. 7, AT THE END (f), AND PERFORMING A CIRCUIT, RETURNS INTO THE SAME BODY AT (g), AND A CONTRARY AND EQUAL EFLUX, CIRCUIT AND INFLUX, PROCEEDING FROM (g), AND RETURNING AT (f), PRESERVES THE BODY, ACTUATED AT ONCE BY THESE CONTRARY POWERS, AT REST, AND A BODY (n) BE PLACED WITHIN THE EFLUX OF THE FLUID AT (f), AND THIS EFLUX ONLY, ACTS UPON (n), THE BODIES M AND (n) WILL MUTUALLY RECEDE FROM EACH OTHER, WITH EQUAL MOMENTS.

For the body (n), by receiving the action of any portion of the circulating fluid, deprives the
body

body M of the action of that part, and thereby deftroys the equilibrium of M, by diminifhing, on one fide only, the furrounding activity. The iffuing fluid from (f) remaining in (n), is prevented from performing its circuit and returning at (g); on the fide (g), therefore, the preffure on the body M is diminifhed, by fo much of the action as (n) fuftains; while, on the fide (f), the other current (not defcribed in the figure), proceeding from (g), enters undiminifhed; the body M, therefore, being preffed as ufual on the fide (f), and having part of its equal preffure at (g) removed, will begin to move towards (g), retiring from (n); while (n) retires alfo from (n), by the out-flowing current which it arrefts, and the current which moves (n), being the fame, which is prevented from entering M at (g), thefe contrary motions muft, neceffarily, be equal.

Observation VIII.

The phenomenon of repulfion, correfponds L with the foregoing propofition and explanation. If fimilar poles, North or South, of a magnet, are brought near together, they will recede from each other, with equal moments.

Explanation.

An active influence, furrounding the magnet, M acts at a diftance from it. If a body be rejected from the magnet, this influence is directed alfo from the magnet; if the influence iffues continually, the fource muft be fupplied: the moft fimple, obvious, and natural fupply, is a circulation of the fame influence; if the iffuing influence be found to recurvate backwards, towards the end, where the fupply fhould enter, and not to proceed continually from the magnet, its return into it is put out of doubt. That the influence does fo recurvate, towards the contrary end of the magnet, is made to appear, by ftrewing fteel filings between two repellent ends of two magnets, placed near together: that the repelled body is actuated by portions of this out-flowing fubftance, demonftrates that the influx, at the contrary end, is diminifhed; for the portion that is employed in repelling the body, cannot, at the fame time, be purfuing its courfe in the circulation: this fact, therefore, may be confidered as completely in point to the propofition.

ALL ACTIONS HAVE A RELATION TO TWO BODIES, WHICH, BY THE ACTION, SUFFER CHANGES, EQUAL IN CONTRARY DIRECTIONS.

OBSERVATION IX.

The fact is univerfally eftablifhed by experiments and obfervations, and needs no particular example here.

This

This propofition is not to be underftood, as being the fame with the third law of motion in the received theory, *that action and re-action are always equal and contrary.* This law appears, from what has been faid in the four preceding propofitions, to be an erroneous hypothefis to explain a general fact.

The changes, in all action, are equal and contrary in two bodies, but this does not proceed from a re-action to every action, or require that all action fhall be mutual and contrary. In all cafes, a fingle action, without any re-action, or mutual actions, ferves to explain the phenomena; while the hypothefis of re-action or mutual action, if admitted, is contradictory, even to appearances.

By recurring to the preceding illuftrations, (z 291, C, D, F, K, &c.) it will appear, that in all action, whether impulfe of bodies, or the action of furrounding fluids, the body acted on is a recipient of an active fluid; that another body is the fource of that fluid, and is deprived of fo much as it has imparted to the former; there is, therefore, in *one action* the *two changes,* equal and contrary, which the fact exhibits. If a fluid iffues from a fource into a receptacle, the former is exhaufted, as much as the latter is replenifhed;

the action is one, but the change effects two objects, in a contrary manner, and in an equal degree.

s   In some cases, re-action is impossible, in fact, and inconsistent with the appearances; it is always unneceffary, and in no inftance proved.

t   In all impulse, it is impossible that any action on a body should give birth to its own contrary and destroyer—all nature revolts at this idea, and all analogy contradicts it. It is inconsistent with the inactivity of body. If inactive in itself, it can act only by an acquired activity; but a re-action, contrary to the action acquired, must originate in the body itself, contrary to its alleged inactivity. It is inconsistent with the communication of motion; for, if this law were true, all motion would be destroyed in impulse and none could be communicated; there would be an universal stagnation and repose wherever bodies met or impelled.

u   Such would be the consequence, if this law, celebrated so highly in the writings of philosophers, had any place in nature. The contrary changes from whence a re-action is inferred, afford the proof against it, unless that self-evident law be denied, that contrary and equal

actions

actions meeting, destroy each other; for if the actions be destroyed, they can produce no changes.

In actions, like those termed attractive, be- x cause both bodies are moved, both are concluded to act; but this inference is not just, since it has been proved, that one action is sufficient to the effect, and, therefore, two ought not to be admitted.

There are some familiar facts adduced in sup- Y port of this law, which it would be inexcusable not to mention.

" If you press a stone with your finger, the z " finger is also pressed by the stone." " If a " horse draws a stone tied to a rope, the horse " (if I may so say) will be equally drawn back " towards the stone." Here, one can only say, it is astonishing how far an hypothesis can obscure the clearest views of reason. Both the preceding assertions, are contrary to the clearest evidence, yet both meet a ready acceptance with the majority of men. The stone presses not the finger, nor does it draw back the horse. In what is the animal structure superior to a stone, if the latter can return the action the former exercises? How do these actions appear?

Instead

Inſtead of a ſtone, let one finger preſs another, which returns the preſſure; let the horſe be oppoſed by another horſe. The obſtruction to motion will then evince the contrary actions—friction may, in like manner, prove an obſtacle—but let the ſtone be free, it will obey the impulſes of the finger and the horſe, and thereby prove that it re-acts not. The finger feels only its own preſſure; the rope tied to the ſtone is diſtended only by the action of the horſe; but if, as it is ſaid, the horſe is drawn to the ſtone, as much as the ſtone to the horſe, ſurely the ſtone may as reaſonably be expected to walk on and drag the horſe after it, as the horſe to get the better of the ſtone, theſe opponents being declared, in all reſpects, equal.

A When in the Scholium, at the end of the firſt book[a], we find Sir Iſaac expreſſing himſelf thus: " I was only willing, by theſe examples, to ſhow " the great extent and certainty of the third law " of motion:" we are not to imagine, that ſo accurate an experimentaliſt raſhly pronounced, concerning a fact which he had not duly aſcertained. He has elſewhere expreſſed the fact unexceptionably, " that equal changes, were in " his experiments always produced towards con-" trary parts." Of the changes, the ſenſes and

[a] Principia.

admeaſure-

admeasurements were the criterions; but his consequence, immediately drawn, is erroneous, " that " the action and re-action were always equal." Here he assumes that in order to two contrary changes, two contrary actions are necessary; but we have shown that in the actions of a medium between distant bodies, these are unnecessary, and in impulse, where, if the contrary actions met, they would destroy each other they are impossible.

IF A BODY IN MOTION WERE EXPOSED TO NO OBSTACLE, AND SUBJECTED TO NO NEW INFLUENCE, IT WOULD CONTINUE IN ONE UNIFORM RECTILINEAL MOTION. B

We have no fact in direct proof of this proposition; all bodies, that we can observe in motion, are both exposed to obstacles and subjected to new influences; but inasmuch as bodies are disposed, to continue in an uniform motion, and to move in right lines, except so far as new influences and obstacles appear, facts, may be said to correspond with the proposition.

The first law of Sir Isaac Newton expresses the rectilineal tendency of motion, and the perseverance in motion, of bodies, once moved. But the cause, and the conditions are different from those here assigned. An innate force is the cause. C caufe; the conditions are, unlefs any new force produce a change. When a body is moved, a force or active fubftance is imparted to it; and the body continues to move, unlefs the efficacy of this force be counteracted by a contrary force imparted, or the body be deprived of the force, by its own impulfe on other bodies; two caufes diminifh or deftroy motion; a new and contrary force, fuffered, or the force of the body in motion exerted on fome other: a new force impreffed on the moving body, is not, therefore, the *only condition* of change in its motion, as the firft law of Newton declares, nor is an *innate force* the caufe of its perfeverance.

D   A BODY IN MOTION, MEETING WITH A-NOTHER BODY IN ITS PATH, WILL COMMUNICATE MOTION TO THAT OTHER.

The body is moved by a fubtile fluid, which, having motion effential to itfelf, has been accidentally imparted to the body, which, with the fluid, acquires its motive tendency. Any other body in the path of fuch a motive body, and not difpofed to move with it, muft be an obftacle in its way, by reafon of their mutual refiftence to penetration; but the fubtile fluid, incapable of reft, or having its effential motion deftroyed, and eafily pervading all bodies, paffes on, out of the former motive body, into the obftacle, to which

it

it imparts its motion, and thus, in impulse, motion may be said to be communicated.

The fact of communication of motion, here deduced from principles, being on the received system inexplicable, is there considered as a miracle, or an effect not depending on any natural powers or laws; but what can more strongly evince the imperfection of principles intended to explain natural appearances, than their failing in regard to a fact, of all others, the most frequent and familiar.

THE EFFECT OF A BODY IN MOTION, EXERTED TO OVERCOME ANY UNIFORM FORCE OF GRAVITY, OR COHESION, IS PROPORTIONAL TO THE MASS COMPOUNDED WITH THE SQUARE OF THE VELOCITY; ALTHOUGH THE FORCE OR MOMENTUM IS IN THE RATIO ONLY OF THE MASS AND SIMPLY THE VELOCITY.

Almost infinite has been the perplexity, which this principle has occasioned, since Mr. Leibnitz first announced it, but in too general a way, and the followers of 'Sir Isaac Newton contended against it. The experiment is easily made.

OBSERVATION X.

Let a weight fall from a known height perpendicularly upon a nail, whose point is just in-

serted

ferted in a piece of wood, and the number of ftrokes be found, which the weight at that height requires to drive home the nail. If the weight then be let fall from a height double the former, one fourth the number of blows will drive home a nail of equal fize and figure, in the fame wood.

H The fact is an obvious confequence of the principles by which action is explained. It is evident, that a projectile impelled againſt the flow of a current, will with a double velocity proceed four times the diftance; for with the double velocity, it having a double motion, it will, by its double motion, be able to encounter a double quantity of the current. But by reafon of the velocity being double, it will have proceeded double the diſtance of the former in an equal time, and in this equal time, it can have encountered a force of the current, equal only to that which the former projection had met in the fame time and half the diftance; for we do not here fpeak of the quantity of the mafs of the fluent, but of its flowing impetus, which being fuppofed uniform, muft always be equal in equal times, without any regard to the mafs or diftance in which the impetus is confidered. At the end of the double diftance, therefore, the fecond projection will have met as much of

the

the impetus as the former met, and as much of its force will be deſtroyed as is equal to the former projection; but the force of the ſecond projection being double, will yet proceed on to another diſtance, alſo double the diſtance of the firſt projection, and thus the whole diſtance of the ſecond projectile will be quadruple the firſt.

It has been proved, that all action conſiſts in motion, and is, therefore, as a current, to have the time of its flow regarded in its meaſures. It is half in half the time, in double the time, double.

While an increaſed velocity augments proportionally the force of a projectile, it diminiſhes in the ſame ratio the reſiſtence of the oppoſing force. Leibnitz miſtook, in ſaying the force of the body was increaſed according to the ſquare of the velocity. Its effect only is increaſed in that proportion, by the diminution of the reſiſtence; the force of the body being in the ſimple ratio only of the velocity, as the Newtonians juſtly ſay. One can hardly have any ſingle evidence of the truth of principles, more ſtriking, than their ſerving to demonſtrate, *a priori*, a fact, ſo important in mechanics, yet ſo perplexing to philoſophers as this has been.

CHAP. III.

Agreement of Phenomena, with the Orbicular or Revolving Motion, of the ACTIVE SUBSTANCE, *which is the Conſtructive Principle throughout Nature.*

A   THE ACTIVE SUBSTANCE, the primary fluid, the immaterial baſis of all material being, the *matter*, if I may ſo ſay, or component ſubſtance of the world, is conſtituted into the fabric of the univerſe by peculiar motions, in orbits, either in circles or in ellipſes of various forms.

B   The truth òf this has been placed out of doubt, by the cleareſt evidence of rational proofs; but it muſt be confeſſed to be a deſirable thing, to be able alſo, to afford ſome teſtimony of which the ſenſes may judge.

C   Matter is an inactive being, which when immerſed in ACTIVE SUBSTANCE, will be moved

by

by its motion; unlefs it be at once actuated equally, in contrary ways.

If, therefore, the ACTIVE SUBSTANCE does, D when from contrary points it is determined to a centre, revolve in fome orbit about that centre, this revolution ought to carry about, in the fame orbit, fuch bodies as are the feats of the revolving ACTIVE SUBSTANCE, if thofe bodies are free to move, and not urged equally, and, at the fame time, in contrary directions.

If a folid body, be at once preffed by other E folids, in two, or in any number of contrary directions, there is, from every fource of the preffure, a flow of the ACTIVE SUBSTANCE into the fource of the oppofite preffure, or a continued circulation produced by the contrary preffures (P 260). But in this cafe, each body is the feat of contrary currents, and by thefe contrary currents, no one of them ought to be moved; therefore, in the mechanical preffure of folid bodies, there can be no fenfible motion of the bodies to evince the internal circulation of the active fluid.

For every preffure has its own diftinct circu- F lation, and the central body, preffed by furrounding bodies, by that preffure, coheres,
becomes

becomes one body with them, and cannot therefore move.

G   If a portion of the active fluid be determined to unite in a centre within itself, there will be infinite currents concentrating as radii from the circumference to the centre, and in the fluid portion, all thefe currents will combine to give the whole portion one revolving motion about its centre, beginning at the centre, and extending towards the circumference.

H   If a material fluid, air or water, be made to move from a circumference to a centre, the progrefs and action of the active fluid will be marked to the eye; for the concentrating streams of the material and inactive fluid, are fo many reprefentatives and fenfible tokens of the invifible fluid, which actuates and moves them.

I   The parts of the fluid do not cohere, but are capable, in any part within the circumference, to receive a motion feparate from the motion of other parts; nor is the fluid, preffed on all fides to the centre, one body, united with and held to the preffing bodies, as a folid preffed on all fides is: a fluid, therefore, is a fit fubject to exhibit to the fenfes as a fenfible covering,

ing [a], the fact of the revolution of the invisible active fluid, and in all fluids and bodies immersed in them the fact ought to appear.

Hence we may lay down the following propositions. K

IF THE ACTIVE SUBSTANCE BE, ACCORDING L TO THE CONCLUSIONS ALREADY DRAWN, MADE TO ASSUME A REVOLVING OR AN ORBICULAR MOTION ABOUT A CENTRE, IN WHICH ALL ITS PARTS ARE DISPOSED TO UNITE, ANY PORTION OF WATER, AIR, OR OTHER FLUID, WILL ALSO REVOLVE ABOUT ANY CENTRE WITHIN THE SAME PORTION, TO WHICH ALL THE PARTS ARE MADE TO MOVE:

And converfely,

IF A PORTION OF ANY MATERIAL FLUID, M ALL WHOSE PARTS ARE MOVED TOWARDS A COMMON CENTRE, DOES FROM THAT CENTRE, TOWARDS THE CIRCUMFERENCE, BEGIN TO REVOLVE ABOUT THE SAID CENTRE OF APPROACH, IT IS THENCE TO BE INFERRED, THAT AS THE MOTION OF THE MATERIAL

[a] The immaterial fluid is embodied in matter, and is evident to our fenses, not directly, but through the medium of its corporeal companion.

FLUID

FLUID TOWARDS THE CENTRE, WAS PRODUCED BY A SIMILAR MOTION OF THE INVISIBLE AGENT, SO, IN LIKE MANNER, THE CIRCULAR OR REVOLVING MOTION OF THE MATERIAL FLUID, DOES DEPEND ON A SIMILAR MOTION IN THE IMMATERIAL.

OBSERVATION XI.

N   If, at the bottom of a veffel of water, an aperture be made for the fluid to efcape, it will revolve about the aperture, at, and at fome diftance from it, and efcape with this revolving motion.

The water rufhes from all fides, in concentrating ftreams, to fupply the continual wafte at the aperture: thefe currents of water indicate the fame central motions of the invifible ACTIVE SUBSTANCE, and the revolving motion is alfo a confequence of the revolution of the ACTIVE SUBSTANCE fucceeding to its central motion.

OBSERVATION XII.

O   If a fluid be agitated in a veffel, by a body immerfed therein, fo as to render the furface uneven, and to produce currents, in various dictions, feveral vortices will appear in the fluid.

The

The fluid rushes centrically from the higher parts to the lower, to restore the equilibrium of the surface; or, streams meet recurrent streams; in either case, producing the effect, on the same principles, as in the last phenomenon. We cannot stir our tea without observing this fact; in every stream it meets our eye; in large bodies of water, or the sea, encountring currents produce whirlpools, the terror sometimes of navigators.

Observation XIII.

Concentrating currents of air rushing from opposite points, produce a revolving motion about the central point, or, a whirlwind.

It is agreed, among naturalists, that whirlwinds happen only where contrary winds meet from several points; they are found of various magnitudes and violence, according to the quantity and velocity of the central winds; at sea, they form water-spouts; on shore, whirlwinds are sometimes alarming and mischievous; in windy weather we may frequently observe small whirlwinds wherever obstacles interrupt the regular currents of air and produce eddy winds, dust and light bodies are carried round in small circles, exhibiting in these familiar appearances, the universal law and constitution of nature.

Q   A body, floating in any fluid, obeys the motions of the fluid, and is, in this refpect, the fame with a portion of the fluid equal to that which the floating body difplaces; a body placed in the centre of any vortex of water, will be made to rotate on its axis; and if placed within the vortex, at a diftance from the centre, will be carried about the centre together with the fluid. Thefe facts may eafily be brought into view by fmall vortices of water, fuch as are found in currents, or as may in various ways be produced in veffels (N, o 312), and may be diverfified at pleafure, fo as to exhibit fome of the moft important and interefting phenomena in nature.

OBSERVATION XIV.

R   The motion of the earth about its axis, being affumed as fact, is a fublime and magnificent teftimonial to our purpofe.

S   For the weight of bodies on its furface preffing from every part towards its centre, does by our principles, afford the inference of a circulating fluid within; and the whole earth, with its atmofphere, whether, as by our theory, it be fuppofed immerfed in the centre of a fluid vortex, or as by the received fyftem, to be furrounded only by a void fpace, it is, on either

fup-

fuppofition, equally at liberty to move and rotate on its axis; it ought, therefore, by the circumambient preffure, which manifeftly exifts, to rotate on its axis, there being no impediment to fuch motion; the fame would obtain in fmaller folid bodies on the earth's furface, when preffed by furrounding folids, but that by the preffure of the furrounding folids they become fixed, and united to them. Thofe bodies alone, whofe external furfaces are at liberty, can be made to rotate by the concentrating tendencies of all their parts. Such alone are bodies furrounded with fluids, or with a void, thefe are found in all cafes, actually to have fuch a rotation; and the proof, fo far as facts can go, is unequivocal and complete.

May we not juftly draw from the above, a new and cogent argument for the truth of a theory which leads, *a priori*, to a demonftration of the earth's motion on its axis. This important fact, hitherto known, only, as the flow refult of aftronomical difcoveries, and, through fo many ages, concealed from mankind, is a fimple and obvious deduction from a juft theory of mechanical motion.

Nor would it be difficult to extend the fame method of deduction further into the machinery of the celeſtial bodies, but it would be foreign from our preſent buſineſs.

x   And to thoſe, who would withhold their aſſent to the *proof* offered of the principle, we may now preſent to their ſenſes the ſame propoſition as the aſſertion of a *fact*, evident to all men.

y   I beg, however, that this appeal, on my part, to facts, may not be miſconſtrued into an acknowledgement that facts are proofs of truths, or that any *propoſition* can directly be proved by *facts*. A proof is a ſucceſſion of conſequences drawn from a truth admitted, ſhowing it to involve another truth, not admitted, until ſo proved. Theſe can exiſt only in the mind, and their ſubject is only general ideas. Every fact is particular, and exiſts without the mind. It is an individual exiſtence, which involves no other exiſtence than itſelf: more eſpecially a fact ſeen, can be no proof of another fact, unſeen, and unobnoxious to any ſenſe. Since then, the cauſes of phenomena are not the phenomena themſelves, they muſt be exiſtences not to be traced by corporeal organs, but by the intellect alone.

The facts above referred to, are ſenſible appearances of vortices in ſenſible fluids, water or air. Theſe facts are no proofs of other facts, they

they show nothing beyond themselves: when, therefore, we bring these as evidences to support other facts, to wit, similar, but invisible motions of an invisible fluid; that evidence is not contained in the facts adduced, but in a process of the mind, which, admitting the fact seen, as an axiom or first principle, infers something further, as necessary from the axiom, admitted. It infers from the effect, a cause adequate and present, and that can be no other than an agent, present in the material fluid, moving invisibly, as the material fluid is seen to move.

But it is too much the wisdom of the present age, to confine themselves to coporeal methods of studying nature. Experiment, alone, claims the philosopher's regard. Having observed the ill success of our forefathers, who, in their closets, abstracted, and in profound meditation, endeavoured, by various processes of reasoning, to come at a knowledge of the secret causes of those things, which the vulgar see with inattentives eyes; and concluding from their failure, not merely that they had reasoned wrong, but that they were wrong in expecting that reason would avail them in this pursuit: the moderns attempt a different method, instead of their thoughts and their pens, they employ their hands and their machines. To the more common facts,

facts, they add a numerous lift of others, lefs familiar, gathered from a variety of arbitrary affociations and conjunctions of bodies, which the pleafure of the operator fuggefts. Thefe obfervations being made with inftruments purpofely calculated, the quantities of the refults can often be afcertained with confiderable accuracy, and from hence much utility may accrue. But in all this, I can fee no other hope than that of new facts continually multiplied by the indefatigable experimentalift. I fee no advance towards the knowledge of caufes. I cannot conceive how the caufe of common appearances can be expected to be found in other appearances, lefs common and equally, or more obfcure. I can difcern herein frefh labour, only, cut out for the philofopher, but no affiftance towards the attainment of his end.

I confefs I am aftonifhed, when I fee philofophers of modern times pouring contempt on the ancients for the folly of hoping that reafon would arrive at a difcovery of the myfteries of nature, and commending their own fuperior wifdom in the happier method they have adopted, when I confider what this method is.

That in the place of reafon, they employ mechanical inftruments, and fubftitute handicraft working for thinking.

Are

Are the secrets of nature remote, that telescopes can bring them to us; are they small, that microscopes can discover them in their retirements; are they pent up in the innermost parts of hard bodies, that, hammar, chissel, and saw, can break their prison walls, and let the captives free?

By force of mechanism, by tortures, do we A hope to drag from nature, the secret which she refuses to our prayers.

After all, what has experiments done for us, B more than increase the catalogue of facts. Do we know more of causes than our predecessors? What, after all, can we expect? Will the cause we seek assume a corporeal form, and charm our senses with unknown delights: will it be beauty to the eye, music to the ear, or sweetness to the taste; may we handle it; will the mathematician determine its figure; can we *keep* it; shall we treasure up the precious novelty with the regalia of state? And should we, by some lucky hit, some exquisite jumble of matter and motion, arrive at this desired end, how long should we be, before every branch of knowledge would adopt the experimenting plan! The mathematician would seek his *definitions* and *axioms* in the collisions of bodies, and leave his *demonstrations*

ſtrations for the more *certain* way of rule and compaſs; clocks and watches would be conſtructed to point out the moral relations and duties of mankind.

We ſhould ſeek a knowledge of the *firſt cauſe* in the ſame experimental way, and in the ſchool of nature (as the moderns ſay) ſtudy natures God. Some would ſeek him in the rude and boiſterous elements; others in the mechanic's ſhop, where the fetters of art lend their aid to ſubdue nature; the poliſhed mirror, the pervious lens, the wheeled machine, the force of mechanic powers would all be ſet to work in the pious cauſe. The chemical theologiſt by more ſubtile means, by alembic and retort, would ſeek to explore the latent deity.

c I deſire not to abate the zeal for acquiring facts, nor decry their utility in their proper place. It is againſt the relinquiſhing for this, all other knowledge, and all other purſuits, that I proteſt; againſt the miſtaken zeal that explodes and brands with opprobrium, under the name of *metaphyſical*, all thoſe inquiries that alone can give utility to experimental knowledge. And when I ſee thoſe who call themſelves philoſophers, exulting in their error, glorying in their ignorance of that which only can be called philoſophy, and in that their enlarged knowledge of facts,

facts, which but serves to make the want of a just theory more conspicuous; deceiving, thus, themselves and posterity; I hold it my duty, as a good citizen, and the friend of mankind, to declare these sentiments without reserve; nor to a false delicacy, sacrifice the most important interests of truth and of the community. Neither in this declaration of my firm persuasion, can I allow that numbers, authority, or talents, are against me; the fashion, indeed, is: but if to the many respectable names in the present day, who are convinced of its errors, I add the weight of all antiquity, I may hope to stand excused of arrogance or presumption in joining my voice to so great a majority of the world, against the fashion, however respectable, of a single century.

CHAP. IV.

Select Propositions, abridged from the foregoing Pages, and separated from their Proofs.

THE following selection of propositions from the foregoing work, as there maintained, with general references to the parts where they are to be found, may be useful, in some respect, as an index to the principal parts; but chiefly as exhibiting, in one view, the substance of the opinions, with their order, whereby the reader, who has gone through the arguments severally, may have a collected view of the doctrines, apart from their proofs, and thus be more easily possessed of the substance of the work, and more competently, as well as more satisfactorily to himself, form his judgment upon the whole.

PROPOSITIONS.

A Sensations are the original sources of all our knowledge (H 4).

First

First principles exist in nature, and only in B nature are to be fought (1 4).

Being concealed among more obvious quali- C ties, they are to be fought by analyfing those things wherein they exist (L 5).

These are our fenfations derived from the D existences about us (A 322). It is, therefore, by the analyfis of our fenfations, we are to feek natural principles (M 5).

First principles are not objects of fenfe, E therefore, not to be fought by experiment, but by the intellect alone (N 5).

Principles obtained by reafon are equally F certain with facts (O 6).

The fecondary qualities of bodies are re- G folvable into the idea of power (P 11).

SOLIDITY ANALYZED, RESOLVES ITSELF H INTO POWER OR ACTIVITY.

Inactivity is a privation, and incapable of I analyfis.

K   Matter is both active and inactive in different respects; its component parts are active, and by their action constitute matter as an whole inactive (M 18).

L   Inactivity has no relation to motion, but is its negation (O 18).

M   THE ESSENCE OF BODY IS POWER (U 25).

N   COHESION IS A POWER, WHICH, EITHER IN HOLDING TOGETHER ANY NUMBER OF PRIMARY SOLID PARTS, OR IN CONSTITUTING ONE PRIMARY SOLID, BY HOLDING ITS OWN PARTS TOGETHER, IS SIMILAR, BOTH IN ITS NATURE AND MODE OF OPERATION (N 25).

O   The power of solidity is twofold, from within, outward, resisting compressure; from without, inward, preventing expansion (E 29).

P   MATTER IS NOT IMPENETRABLE (Y 37, &c.)

Q   THE ANALYSIS OF THE IDEA OF MOTION RESOLVES IT INTO ACTION ALONE (P 48).

R   ACTION DEPENDS UPON AND IMPLIES THE PRESENCE OF A SUBSTANCE WHICH IS ACTIVE (A 57).

AN

AN ACTIVE SUBSTANCE EXISTS THROUGH- S
OUT NATURE, THE UNIVERSAL ESSENCE AND
AGENT (s 61).

This ACTIVE SUBSTANCE does actually exist, T
external to our minds, and independent on being
perceived (Chap. vi. p 63).

The ACTIVE SUBSTANCE is immaterial and U
unintelligent, intermediate to matter and mind
(Chap. vii. p 84).

The standard of truth in philosophy is the X
mind's perception of the connections of things
(Q 98); hence rules of reasoning (A 102).

THE MANNER IN WHICH THE ACTIVE SUB- Y
STANCE IS ACTIVE IS BY MOTION (O 106).

THE MANNER IN WHICH IT ACTS IS BY Z
UNION WITH THE SUBJECT OF ITS ACTION
(T 107).

NO RATIONAL AGENCY IS DISCOVERABLE, A
IMMEDIATELY IN EXPERIMENTS, BUT BY IN-
FERENCES FROM THEM (P 112).

BODIES AS ENTIRE MASSES BEING INACTIVE, B
THEIR ACTIVITIES ARE ADVENTITIOUS, NOT
INNATE (A 113).

They

c   They have two forms of activity, motion and impulse (B 113).

D   They are active by means of an ACTIVE SUBSTANCE communicated to them (H 115).

E   THIS ACTIVE SUBSTANCE PENETRATES THE SOLID PARTS OF BODIES (M. 116).

F   The quantity of activity is proportional to the quantity of ACTIVE SUBSTANCE (Q 117).

G   IMPULSE DEPENDS ON ACTIVE SUBSTANCE FLOWING OUT OF THE IMPELLING BODY (R 127).

H   MOTION DEPENDS ON THE ACTIVE SUBSTANCE RETAINED WITHIN THE BODY IT MOVES (S 128).

I   ACTIVITY IS COMMUNICATED IN IMPULSE BY THE FLOW (G 326).

K   Percussion is a mode of impulse (I 326), marked by the change from the retained state of the ACTIVE SUBSTANCE (H 326) to its flow, or it is the beginning of pressure, or the mean state between motion and pressure (A 144).

L   Pressure is the last and complete state of impulse, and is when the flowing state of the ACTIVE SUBSTANCE completely exists (B 144).

OUT

OUT OF THE IMMATERIAL ACTIVE SUB- M
STANCE MATTER MAY BE FORMED (C, D 148).

Matter being at once active and inert in N
different respects (K 324) has, by its activity,
a relation to immaterial active natures (F 149).

The union of body and mind, is a fact in O
evidence of their mutual relations (X 156).

Solidity confists not in an inactive fulnefs P
(G 159).

Denfity depends not on porofity (I 160). Q

A MATERIAL ATOM IS FORMED BY THE RO- R
TATION ABOUT ITS OWN AXIS OF A PORTION
OF THE IMMATERIAL ACTIVE SUBSTANCE
(C 166).

The two effects, folidity and inertia, arife from S
one caufe, a rotatory motion. In the modern
fyftem, one effect motion, requires two caufes,
the impreffed and innate forces (N 170).

The relation between the ACTIVE SUBSTANCE T
and matter is refembled to that between a line
and a circle (O 171).

The

u   The modern philofophy errs in attempting to prove powers or exiftences by mathematical reafoning (z 173, c 174).

x   It is an error to imagine that the Newtonian fyftem is founded upon mathematical reafoning, or fupported by it (c 174).

y   Its phyfical principles (or metaphyfical) are incapable of fupport, but in their own merits (d 175).

z   PROBLEM PROPOSED TO EXPLAIN THE MANNER OF THE ATTACHMENT AMONG ATOMS (G 175).

a   The difficulty which hitherto has attended this inquiry, has been of our own creation, by the falfe premifes affumed (1 176, &c.).

b   ATOMS ARE UNITED, OR COHERE BY INTERMIXTURE OF THEIR SUBSTANCES (A 183).

c   THERE IS A SIMILAR LAW OF CONSTRUCTION, OR, AN ANALOGY BETWEEN ALL PARTS OF NATURE, THE SMALLEST AND THE LARGEST (Chap. vii).

Bodies contain within them active
principles, and sources of action, although they are inactive (p 228).

Motion is the original state of being
(L 229).

Rest is derived from motion (m 229).

The orbicular motion, which conftitutes matter and the rarer orbs, that retain the planets and
connect the parts of the univerfe, arife from a
law determining the motions of the ACTIVE
SUBSTANCE to union, and from the neceffity
of exiftence (Chap. iii. p 233).

Cohefion is produced by imperfect atoms,
called *connecting corpufcles*, of the fame nature
with the primary atoms, compounded of parts
of the atoms they ferve to connect (Chap. iv.
p 247).

Mutual contary impulfes are tranfient unions
produced by the ACTIVE SUBSTANCE, by means
of its orbicular motion, whereby it produces all
unions (Chap. v. p 257).

The queftion of the origin of motion implies
an abfurdity; that death preceded and produced life (Chap. vi. p 263).

An important difference between the law of union and attraction, is, that one is said to be imposed upon an immaterial active fluid, the other on, inactive matter, which latter is impossible, as it implies the same quality, activity, to be and not to be, at the same time.

CONCLUSION.

CONCLUSION.

HERE, for the prefent, I am compelled to clofe a long and interefting labour; under many difadvantages, I fubmit a faulty, and unfinifhed work to the public award. Fully fenfible how much I ftand in need of indulgence, I am yet apprehenfive, fome may think me leaft entitled to it, in a point, where I reft my chief pretentions to favour; I mean, in having given my own reflections, freely, without fuffering them to be influenced, by what the world calls authority.

What will be the fate of this volume, time, only, can reveal; of this I am unalterably affured, every free difcuffion of eftablifhed doctrines, produces a double benefit to fcience; as it afferts the independence of the mind, and tends to diffolve the charm which eftablifhments, long unexamined, produce; while it continually brings them anew to the teft of reafon, and gives us the chance of difcovering errors, which our predeceffors may have overlooked.

Divine revelation, alone, can have a claim to an independence on reason. Whether those are worse philosophers, or theologists, I know not, who would place philosophy on the same footing.

Nor is it admissible, as these contend, that *what has once been proved to be true, must ever be so*; designing, thus, to persuade men, that none need again examine what *they* have reported as true. For what is proof? It is that which satisfies the mind of truth. It is an evidence adapted to the state of the intellect. The same thing will, to one man, be demonstration, and give full assurance, which, to another will not amount to probability. Would not the followers of Aristotle, or of Des Cartes, aver, they have proofs of their doctrines; and if what has once been proved true, must ever be true, our system, each would say, must be eternal.

To what then does a principle lead, which will apply alike to every system, and every sect.

The truth is, the votaries of every different philosophy, and different worship, are themselves persuaded in their faith; but this argument can only avail to the predominant party. When Cartesianism was the vogue, would not they have
said

faid what the Newtonians do now. Had the argument been then allowed, the fyftem of Newton would never have prevailed—fhould it be now allowed, Newtonianifm can never be reformed.

Dulnefs will affix the feal of infallibility to the orthodoxy of the day. It will bar the door of fcience upon mankind, and blaft the rifing hopes of all pofterity.

What glorious advances in knowledge, futurity may have in ftore, we know not; but ardent ought to be our hopes, and indefatigable our zeal; we ought to prefs forward, nor imagine the labours of *one man* can have bounded our pro-grefs, nor hearken to thofe, who, having erected an idolatrous temple to *his* fame, would have us *ftop and worfhip*; nor prefume to pafs the boundaries they have marked out to human reafon.

Let thefe boaft of their proofs and enjoy them; let us demand their production, and judge of them for ourfelves. Proofs to paft ages are not proofs to the prefent. As the mind gets enlightened and emancipated, its powers are ftrengthened, its vifual organs emit a brighter ray. In things we deem proved, pofterity may difcern errors. Thus we may go on from age to to age, advancing to a more mature and perfect period than imagination can yet conceive.

Absolute proof, indeed, is ever the same, but of this we have no knowledge. It is only with proofs, *relative* to our own intellects, that we are concerned.

Permit me now to caution those engaged in scenes of active life, from the dangerous error of imagining that metaphysical inquiries; that the first and abstruse principles of things are, to them, of small importance. Arts, manufactures, commerce, laws, government, civil policy, the order and happiness of society, are the result of knowledge and science, must partake in their improvements, and together with them, make advances towards perfection, or sink again into barbarian darkness. The dearest interests of mankind are involved in those inquiries we so little esteem. The fate of posterity, in these respects, is in our hands; we may cherish the seeds from which the rising generation may reap a plentiful harvest of knowledge, or we may choak the soil and leave only barren fields for their inheritance.

The connection, indeed, between metaphysics and the concerns of life, lies too deep to be seen by superficial observers; but the deeper and more concealed is the influence, the more potent and extensive are its effects. A subterraneous expansion of fire, if near the surface, soon makes its harmless exit; but if deep sunk, near the centre, it shakes half the globe, and buries nations in ruins.

Had I foreseen, the fatigue and the inconveniencies to which my present and its consequent occupations have subjected me, and to which they will yet expose me, in one collected view, I am not certain whether my fortitude would have supported me in the undertaking. I have been led on by insensible degrees to the present moment; my desire of information prompted me to think; my wish, both to benefit others, and to profit by their assistance, induced me to communicate my thoughts.

They are now no longer mine, but the propriety of the public and of mankind. If they tend to a growth in truth, I shall rejoice in having contributed to the public treasury of science; if they be the offspring of error, I shall concur in consigning them to oblivion. We should be cautious, equally, neither to hoard up base coin, nor throw away sterling riches;

we

we fhould examine with care, and felect with impartial hand.

As I have fpoken freely of the opinions of others, others are at liberty to do the fame of mine. Although I have fought to detect errors, I have no pretentions to be exempted from erring. But, if I have written with candour, as I have examined with care, the like care and candour others will obferve to me, in their endeavours to correct my faults, which, I hope, all, who have the proper talents, will be more affiduous to amend, than fevere to cenfure.

F I N I S.

Fig. 1.

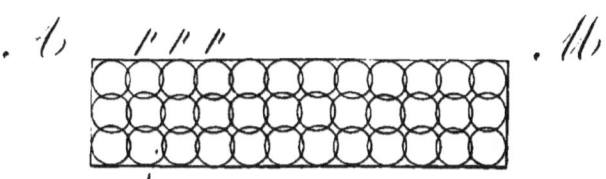

Fig. 2.

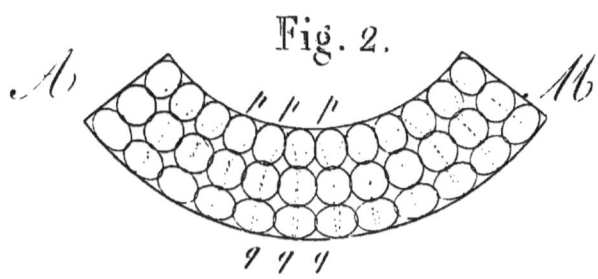

Fig. 3.

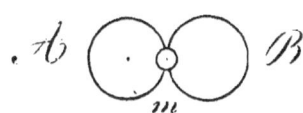

Fig. 4.

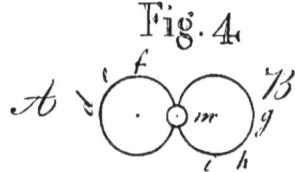

Fig. 5

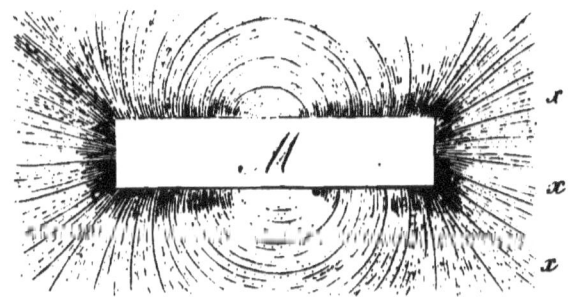

Fig. 6

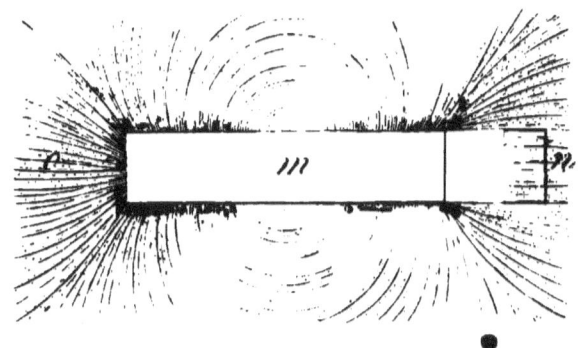

Fig. 7

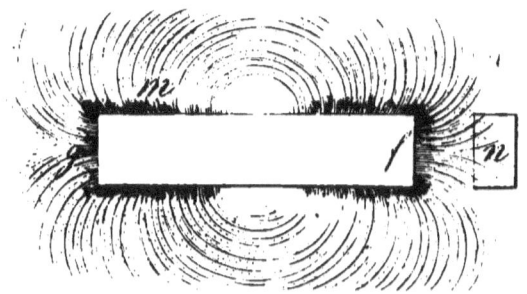

www.ingramcontent.com/pod-product-compliance
Lightning Source LLC
Chambersburg PA
CBHW020230240426
43672CB00006B/481